Simulation Foundations, Methods and Applications

More information about this series at http://www.springer.com/series/10128

David J. Murray-Smith

Testing and Validation of Computer Simulation Models

Principles, Methods and Applications

Springer

David J. Murray-Smith
School of Engineering
University of Glasgow
Glasgow, UK

ISSN 2195-2817 ISSN 2195-2825 (electronic)
Simulation Foundations, Methods and Applications
ISBN 978-3-319-36266-3 ISBN 978-3-319-15099-4 (eBook)
DOI 10.1007/978-3-319-15099-4

Springer Cham Heidelberg New York Dordrecht London

Softcover re-print of the Hardcover 1st edition 2015

Printed on acid-free paper

Springer International Publishing AG Switzerland is part of Springer Science+Business Media (www.springer.com)

Preface

This book is intended to fill a gap in the currently available literature on the development and application of dynamic simulation models. It deals with issues of model quality and, more specifically, with the processes of testing, verification and validation. Since simulation models can never be proved to be "valid" in any absolute sense, the topic of model testing inevitably involves subjective issues and often a trade-off between accuracy, cost and practical issues associated with the intended application of the model. The emphasis within the book is mainly on continuous system simulation problems, and case studies are used to provide examples from the fields of engineering and physiology. The range of these applications and their cross-disciplinary nature reflects my research interests and activities over a period of almost 50 years.

Since the book is aimed at people with interests in simulation models and their use in practical applications in many different fields, some assumptions are made about the prior knowledge of the readers. Relevant supplementary material is therefore being provided through a website (http://www.springer.com/gb/book/9783319150987), and it is hoped that this should provide a convenient way of accessing additional background information, both in terms of the general principles of modelling and the application areas considered in the case studies. For those who do not have a background in engineering and the physical sciences, this includes sections about mathematical and system modelling concepts. Similarly, for those whose prior knowledge is lacking in terms of the biological sciences and who need more in order to understand aspects of some of the physiological case studies, the supplementary material includes sections which present some basic concepts from those areas. No attempt has been made to make the supplementary material sufficient on its own to meet the needs of everyone. Instead, only a brief account of each topic is included on the website, and links are provided to other sources of information which are far more extensive and detailed. The supplementary material also includes some data sets relating to some of the case studies, and it is hoped that these may allow readers to carry out their own investigations of those examples. Frequency-domain and time-domain data from tests carried out on some relatively simple systems and models, which are not discussed within the book, are also provided. It is hoped that these may allow the reader to explore and apply experimental modelling and model testing methods to these additional data sets. All the data sets and models provided through the website may be used freely and shared with others, provided the source is acknowledged.

Since the case studies, and other applications discussed in the book, are drawn from research projects and my teaching activities, I must record my sincere thanks to the many research students, research assistants, undergraduate students and colleagues who contributed in important ways. Some of those receive explicit mention through references to reports, theses and journal or conference publications, but I must express my thanks to all who have contributed to the work in any way. I must also thank students who may have encountered some of these case studies within their courses and whose questions and difficulties have contributed significantly to the way in which material has been presented.

Glasgow, UK David J. Murray-Smith
June 2015

Contents

Abbreviations

ADC	Analogue to digital converter
AGARD	Advisory Group on Aerospace Research and Development (NATO)
AIAA	American Institute of Aeronautics and Astronautics
ANL	Argonne National Laboratory (US)
ASCI	Accelerated Strategic Computing Initiative
ASME	American Society of Mechanical Engineers
BIM	Building Infrastructure Management
CAD	Computer-aided design
CFD	Computational fluid dynamics
DAC	Digital to analogue converter
DAE	Differential algebraic equation
DLR	Deutsches Zentrum für Luft- und Raumfahrt (German Aerospace Centre)
DMSO	Defense Modeling and Simulation Office (US)
DSB	Defense Science Board (US)
ESA	European Space Agency
FMI	Functional Mock-up Interface
FNS	Functional neuromuscular stimulation
FRAC	Frequency response assurance criterion
FRF	Frequency response function
GA	Genetic algorithm
GP	Gaussian process
GP	Genetic programming
ISPOR	International Society for Pharmacoeconomics and Outcomes Research
MC/DC	Modified condition/decision coverage
M&S	Modelling and simulation
M&SCO	Modeling and Simulation Coordination Office (US)
NATO	North Atlantic Treaty Organisation
NSHEB	North of Scotland Hydro-Electric Board
ODE	Ordinary differential equation

ONR	Office of Naval Research (US)
PDE	Partial differential equation
PTB	Project test bed
S3D	Ship-Smart System Design
SA	Simulated annealing
SCS	Society for Computer Simulation
SMDM	Society for Medical Decision Making
SSD	Smart-ship design
TIC	Theil's inequality coefficient
V&V	Verification and validation
VV&A	Verification, validation and accreditation
VTB	Virtual test bed

Chapter 1
An Introduction to Simulation Models and the Modelling Process

1.1 Objectives in Mathematical Modelling and Computer Simulation

Mathematical modelling and computer simulation methods are powerful tools and have applications in many areas of science, engineering, medicine, economics, business and the social sciences. It is clearly possible to describe any real system in different ways and the number of possible models that can be used in any specific case is infinite. In practice, we have to find ways of assessing the suitability or otherwise of a model for a proposed application and for comparing different models in terms of objective measures or, in many cases, through procedures that are more subjective. This book discusses issues of model testing and evaluation, both for engineering applications and for system modelling in the biological sciences. Four case studies are included, two of which are from engineering and two from physiology. The emphasis throughout is primarily on dynamic models that involve variables that are continuous functions of time. The methods being discussed thus relate mainly to models implemented using continuous system simulation tools.

It is important to note, from the outset, that there is an important difference between ways in which mathematical modelling and computer simulation are used by engineers and the ways in which these techniques are employed for broader scientific investigations where the objectives are often very different from those in engineering. Probably the most important factor relates to the uncertainties in our understanding of the real system represented by the model and the extent to which there are unknown, or incompletely understood, elements. Although engineering systems involve uncertainties and models of those systems have limitations, they are often relatively well understood in terms of their structure. Within that field, models can be very useful in specific applications and are most often developed to help in the design of new engineering products, or to allow testing and analysis of

D.J. Murray-Smith, *Testing and Validation of Computer Simulation Models*,
Simulation Foundations, Methods and Applications,
DOI 10.1007/978-3-319-15099-4_1

an existing engineering product or system. Most engineering systems are thus well understood in terms of their structure and are often described as "closed".

In contrast to engineering systems, natural systems arising in scientific fields such as physiology and the environmental sciences often involve models that are not closed. Information about the real system represented by the model is incomplete or has been derived through processes involving fairly drastic simplifications and approximations. The system boundaries are often ill-defined and involve major uncertainties. In science, a model is most often a stepping-stone within a research project which is aimed primarily at providing a better understanding of a natural phenomenon and models may be especially helpful in the design of experiments for the testing of hypotheses. Model development and computer simulation techniques, thus become a central and natural part of the scientific method. It is interesting to note that in clinical medicine we find some applications of modelling and simulation that show quite strong similarities to some types of model-based investigations in engineering, while other medical applications may involve problems which display all the uncertainties and the open-ended nature of investigations in pure science.

In science, observations made of the behaviour of a real system may often be explained in a simple and concise way using a mathematical model or an associated computer-based simulation. More quantitatively, a model may also be used to provide an indirect estimate of something that is difficult to measure directly. Models and the associated computer simulations may be of assistance in making predictions or decisions, such as those relating to climate change, or weather forecasting, or estimates of future changes in air or water quality. They may also have an explanatory role and may be developed as part of an attempt to bring together all the available information about some natural system in a convenient and concise form of description that can be accessed by researchers in different groups around the world.

As with scientific applications, models in engineering may also be used to describe, analyse, explain or simply document a complex system. However, a more important type of application involves the use of these techniques to support the design process and prototype development, or to assist in decision-making processes. Models are often vital for tackling the trade-offs within the design process and properly tested models and computer simulations now provide evidence that is routinely used to establish a basis for certification of the performance, safety and reliability of safety-critical and high-value systems. They provide a way of supplementing the testing of prototype systems and can allow investigation of performance limitations that would not be permitted in more direct ways for reasons of safety or the risk of damage to expensive hardware. Proven models can reduce engineering development times and costs in a significant way and also provide a basis for some techniques of computer-based control, and for more specialised applications such as schemes for automatic fault detection and fault alleviation. Such models are also valuable for the development of real-time simulators that are used routinely for training of operators. Without simulators, risks associated with use of the real hardware would make it impossible to expose operators to training scenarios involving major system problems. A simple example of this could be a pilot being trained to deal with an engine failure or a control surface actuator failure in an aircraft in flight. Other examples of similar safety-critical applications are found in the training of operators for off-shore platforms serving the oil and gas industry, in nuclear power generation and in the control of electricity supply networks.

From all of the above discussion, it is clear that, because of the important role of simulation and modelling techniques in many different fields, the adoption of proper procedures for testing models and computer simulations prior to their routine application is very important. The significance of testing is obvious for engineering design, for training simulator development or for simulators on which different management and operational strategies can be investigated. The testing of models is also critically important in scientific investigations since any publication of results that depend on a simulation model should include details of the processes used for assessing the model's fitness-for-purpose. Publications relating to models and the associated simulation software must provide proper "transparency" so that the reader can extract all the information required to fully understand the model and how it has been tested. Ideally the reader should be able, in principle at least, to assemble the model from the information provided and reproduce all the published findings.

1.2 Requirements Definition and Conceptual Modelling

One key aspect of the model development process is the requirements definition. This starts from basic statements of the purpose of the model, together with statements about its performance, cost and timescale. It ends with a detailed set of specifications and performance targets for the model. Defining the precise purpose of a model often follows on from a functional statement relating to the project for which the model is required and the deliverables from that project. The specification of model fidelity must always be related to the broader performance requirements of the planned application.

Within engineering, a distinction may be made between what have been termed "market pull" projects (perhaps involving design and development of a specific product to meet given performance requirements in a specific period of time) and "technology push" projects intended to assess new areas of technology (such as a new form of control scheme) and reach conclusions about their likely future importance and potential value [1]. In the case of "market pull" projects there is usually a clear problem statement that can be used in the requirements definition for the associated simulation model. With "technology push" projects, on the other hand, the requirements definition for models needed in the investigation may be less precise initially but should always be chosen to be representative of problems to which the new technology could be applied. With technology-push projects a range of different models having distinctly different characteristics in terms of their structure, order, nonlinearity etc. might be needed to allow firm conclusions to be reached about the potential value of the new ideas.

Examples of major "market pull" projects in which modelling failures or inadequacies have led to significant extra costs or to late delivery are well known. For instance, in naval construction in the USA the average over-cost of new ship classes is reported as being of the order of 30 % and it is believed that a significant reason for this is that designs have been released to production prematurely [2]. It has also been alleged that most of the extra costs could have been predicted at the design stage if the systems had been modelled more comprehensively. These ideas form a central part of the reasoning behind the development by the US Office for Naval Research (ONR) of the Ship-Smart System Design (SSD) tool where it has been recommended, in a rather revolutionary set of proposals, that the system model should become the system specification [3].

Although not all projects in all areas of application, even within engineering, can fit within the framework being suggested by ONR, it is obvious that whatever the area of application clear definitions of model requirements are of critical importance. This applies equally to "technology push" projects and to work within other disciplines such as the physical, biological and earth sciences. It should be noted that, in practice, model specifications may change and evolve during a project due to the understanding that is built up during the work, even if the formal requirements remain the same throughout.

Following requirements definition, an early stage of most projects involves the assembling of all available information about the structure and function of the system or the formation of hypotheses for cases involving much uncertainty, as is the case in modelling some physiological systems. The dominant phenomena within the system to be modelled must first be identified and described, initially in terms of words. This could involve energy conversion and storage processes, or material transfer and storage within distinct compartments. Appropriate simplifying assumptions can then be applied to create an initial "conceptual" model. The development of the conceptual model is a highly creative task that often has intuitive elements. The model must include all relevant available knowledge about the important phenomena involved to allow a possible model structure to be defined, together with parameters of the system and important variables. It is particularly important to identify the variables that have measurable counterparts within the real hardware as these are potentially important for model validation.

Essentially, a conceptual model is a collection of statements, assumptions, relationships and data that describe the reality of interest. From this conceptual type of description a mathematical model can eventually be constructed and information useful in the design of experiments to test that model can be derived [4]. Often a top-down approach is adopted, where a relatively coarse type of description is defined initially, with details being added at a later stage. However, there are usually also elements of bottom-up thinking where existing sub-models are introduced within the structure that has been defined in a top-down fashion at the start.

This initial process of conceptual modelling is followed by the abstraction of the information contained within the qualitative description to provide a more formal representation, usually involving the use of equations or graphical block diagram elements. The choice, in this respect, may depend on the modelling tools and computing environment chosen for the work. This mathematical model not only involves equations but also boundary values, initial conditions and data needed to describe the conceptual model in quantitative terms [4]. That can provide information about key parameter sensitivities and inter-dependencies which may be important for design decisions and for performance optimisation.

1.3 Issues of Model Quality

Since a model is only an abstraction of the system it represents, perfect accuracy is impossible. This inevitably raises important philosophical questions but, in all fields in which modelling and simulation techniques are used, the key issue is one of determining the level of model fidelity needed for the intended application. Models also need to be transparent so that all who make use of a model can have some understanding of how it is organised. An inappropriate model is less than useless and, in engineering applications, may delay the project and lead to cost escalation. In scientific research projects, the use of incorrect or poorly understood models may lead investigators in totally the wrong direction. In general, whatever the type of application, modelling errors should be reduced to defined levels for specified operating regions for the system. Information about these modelling errors and a "neighbourhood of validity" must be readily available to users along with all the other information about the model that provide the required overall transparency.

While reducing modelling errors is very important, a balance should also be sought between overall accuracy and other factors. These include development time, solution speed and the cost of developing the model in relation to the expected benefits. In any type of application, the level of detail within a model is linked to its purpose. As models are made more detailed, they inevitably become more complex but model complexity should never be confused with model quality and a simple description can often be better, in terms of quality measures, than a more complex one. Developing a model requires careful examination of information about the real system and consideration of how the model is to be used. In general terms, when modelling a complex dynamic system, it is advisable to move in a stepwise fashion from a well-understood area of operation, such as a steady-state condition, towards situations where knowledge is more limited. Inconsistencies or gaps in the available knowledge can then be found. These may require further experimental work or the testing of an engineering prototype and this process may lead sometimes to a reconsideration of requirements. The outcome of the model assessment process should be a statement of the quantified level of agreement between experimental data and model prediction, as well as information about the predictive accuracy of the model [4].

Tested models allow virtual prototypes to be created before any hardware prototype is available. This provides a way of identifying necessary design alterations at an early stage and, perhaps, of avoiding expensive changes later on. Once real prototypes become available more complete and rigorous processes of testing and model validation become possible, as discussed in Chap. 2.

1.4 Model Re-use

In terms of the overall efficiency of the modelling process, re-use of model components is important and some software tools for modelling and simulation now offer well-documented libraries of re-usable sub-models that are based on the previous experience of many different users. Successful re-use requires sound principles of model management and this is especially important for applications involving large teams of developers, especially when these include multidisciplinary groups and geographically dispersed teams.

In the field of medical decision making some health care models are intended to be "general" in the sense that they can provide a basis for a number of investigations. Other models are built for a single application and are not intended to be re-used but may, in fact, be modified and extended at a later date so that they can be applied to new situations. This division between "general" and "specific" models is also likely to apply in other fields. For a "multi-application" model, with more general applicability, transparency is clearly a priority and the documentation must therefore be of the highest quality. In such cases validation is an on-going process and the model is likely to have to be modified and updated as science advances. In such situations, retaining full documentation for each historical version of a model is important. For a model intended for a single application, issues of transparency and model validation are still vitally important because information about the model has to be fully reported when results from the research are published. Also, a "single-application" model may well be picked up again at some future date by a new user who is interested in a new project with slightly different objectives and may be interested in the possibility of re-using some specific feature of that earlier model.

1.4.1 Model Libraries

A library of models or sub-models, for use in a particular application area, needs not only to be designed to meet current requirements but also to satisfy possible needs in the future [5]. Sub-models should therefore be designed as building blocks for a range of applications rather than specifically for one project. This means that verification and validation processes should be applied, first of all, at the sub-model level and should be subjected to testing over a range of conditions before being accepted, documented and made available for wider use.

One of the most important reasons for model re-use is that it can reduce the time required for the development of new models. A library also allows the investigator to make an informed decision about the sub-model that best meets their needs. This might involve selecting a specific sub-model from a number of representations involving different levels of detail. Within a library, it is useful to establish a taxonomy of models [6], which incorporates generic model classes and sub-classes. This becomes more and more important as the number of models increases. Ideally the library should also allow the modeller to cross from one energy domain to another. As an example, this feature might allow a design engineer to move easily from consideration of a hydraulic actuator for a specific application to examining the possible use of an electrical actuator for the same task.

Some modern object-oriented simulation software environments, such as Modelica® [7], provide standard model libraries and allow new libraries to be developed. Other packages can be extended with tools for physical modelling in various domains. An example of this is MATLAB®/Simulink® [8] which includes some standard library sub-models and, through Simscape™ [9] there are additional standard libraries involving sub-models for fields such as mechanics, hydraulics, electronics, mechanical transmission systems and electrical power systems. Using the Simscape™ language, which is based on MATLAB® [8], new sub-models can be created together with equivalent Simulink® blocks, for components and sub-systems that are not included in existing libraries.

In projects which involve several teams working together, team members may wish to use different tools and languages when building their models of different sub-systems. Sub-models from different software environments may then have to be brought together within some larger model. One example of this is the Virtual Test Bed (VTB) [10] which is an environment that facilitates integration of sub-models developed using other widely-used tools [11], such as MATLAB®/Simulink® [8], Modelica® or VHDL-AMS [12]. This has obvious significance in terms of verification processes and inevitably requires further checks beyond those performed on the original sub-model.

1.4.2 Generic Models

"Generic" models extend the ideas associated with model libraries. A generic simulation model can be applied to a number of different projects without significant internal reorganisation. The essential requirements of a generic description must be identified first and a suitable framework established that offers sufficient flexibility for a number of different sets of objectives. The main benefit of adopting a generic approach is that it may lead to savings in the development of a whole series of models for different projects, compared with the traditional approach involving the separate development of a new model to suit each application.

Benefits may also arise because a generic model requires more rigour in terms of model validation, together with better documentation. However the advantages are only realised if the generic model, once developed, is used for a range of different projects and the potential range of applications requires careful consideration prior to any decision to embark on the development of a model of this kind. Issues arising in the testing and validation of library sub-models and generic models are discussed in Chap. 7.

Examples of the generic approach can be found at present in several application areas, including communication systems (e.g. [13]), automotive engineering (e.g., [14]), electro-optic systems (e.g. [15]) and the planning of critical care resource requirements [16]. A good example is the European Space Agency (ESA) Generic Project Test Bed (PTB) which involves re-usable simulator architectures for spacecraft design [17]. The generic structure includes ground-station models as well as spacecraft sub-systems, together with models relating to the environment. It is important to note that the PTB allows for real-time simulation and hardware-in-the-loop operation and this is a feature that can also be found in some other examples of the generic approach.

1.5 Classes of Model

Many dynamic models used in science and engineering involve variables that are continuous functions of time, such as position, velocity, acceleration, temperature or pressure. Models based on these continuous-variable descriptions may involve ordinary or partial differential equations or differential-algebraic equations. This is the main class of model considered in this book and within this general class there can be many variations in terms of the model structure.

A second class of model that can be important, not only in science and engineering but also in other areas, such as business, planning and operations research, involves discrete-event descriptions. In such models all the variables remain constant between events that mark changes in the model. These changes take place at discrete time instants, either periodically or in a random fashion. Simple examples arise in applications which involve queues, such as in modelling a shop or bank to establish how many tills need to be provided to ensure that customer waiting times are acceptable. A digital computer used for real-time control is another example of a discrete system involving periodic changes. In this case, a continuous variable may be sampled periodically using an analogue-to-digital converter. Calculations carried out using the discrete values obtained from the converter are then changed back into continuous variable form using a digital-to-analogue converter. In modelling this type of component within some larger engineering system we cannot use differential equations because of the discrete nature of the events within the digital processor and an approach involving a discrete model (based on difference-equations instead of differential equations) is more appropriate. However, hybrid models, involving representations that are mainly continuous but do involve some discrete-event elements are becoming increasingly common.

1.5.1 Models Involving Continuous Variables

Within the class of continuous variable dynamic models we can distinguish between models of data and physically-based models of systems. A model of data involves a description fitted to measured responses, usually from a real physical system, leading to a model that expresses an observed relationship between two or more variables. It consists of mathematical functions that may have no direct link to recognisable elements of the real system. Models of this kind are important in fields such as control engineering where input-output descriptions, such as transfer functions, may be used. Often these models may be derived directly from measurements and are often termed "black box" models. They may provide a useful starting point for engineering design but incorporate limited information about internal processes. If they are derived entirely from experimental data their validity is restricted to the conditions that applied in those experiments. Physically-based models, on the other hand, are developed using established scientific principles, such as basic laws and principles from physics, chemistry and biology. The models and sub-models being considered in this book thus range from completely transparent descriptions based on physical principles, through intermediate "grey-box" descriptions, to the entirely empirical black-box form of experimentally-derived model.

Another important distinction is between linear and nonlinear models. Linear models are attractive because they are open to analysis and can be incorporated conveniently into design procedures. However, linear descriptions may be incapable of capturing aspects of the behaviour of the real physical system and issues of nonlinearity should be considered at an early stage in modelling. Assumptions of linearity should not be made without justification and the range of linear operation of the system always needs to be evaluated when a linear description is used. Dangers arise if the model is chosen for reasons of mathematical convenience and the developer fails to recognise properly the complex realities of the real world situation.

A time-invariant description is one in which the performance of the system being modelled is independent of the times at which observations are made. As with questions of linearity, time invariance needs to be demonstrated rather than assumed. Models that are linear and time invariant receive particular attention in many engineering textbooks dealing with topics such as electrical circuit theory, signal processing, dynamics and automatic control. Many systems have properties that allow them to be described by linear time-invariant models for some operating conditions and such models are very attractive because they can be analysed using simple linear methods of mathematics, such as Laplace transform techniques. Although nonlinear and time-varying dynamic models are more general, they are harder to deal with using mathematical methods and numerical and computer simulation techniques have therefore become very important for such models. Simulation thus offers valuable insight for problems that would otherwise be intractable.

1.5.1.1 Models Based on Ordinary Differential Equations and Differential Algebraic Equations

Mathematical descriptions based entirely on linear or nonlinear ordinary differential equations (ODEs) form a particularly important class of model. A broader class of model involves differential algebraic equations (DAEs) which include algebraic relations in addition to the ODEs. In both cases all the quantities in these models are simply functions of time and do not change with spatial coordinates. These are known as lumped-parameter descriptions and are important for many situations involving, for example, mechanical systems, electrical networks and compartmental systems arising in the modelling of chemical processes or physiological systems.

1.5.1.2 Models Based on Partial Differential Equations

Models based on partial differential equations (PDEs) are important for modelling systems that involve quantities that are physically distributed and thus depend on spatial coordinates as well as time. One example could relate to the temperature distributions in a material where, in a lumped representation, this would be modelled in an approximate way by using a mean value of temperature over some region. As well as containing derivatives of variables with respect to time, a model based on PDEs would also contain derivatives with respect to spatial variables. In general terms, lumped models based on ODEs can be viewed as approximations of distributed parameter descriptions based on PDEs.

A simple example of a PDE model is the heat flow equation:

$$\frac{\partial u}{\partial t} - c\left(\frac{\partial^2 u}{\partial x^2} + \frac{\partial^2 u}{\partial y^2}\right) = f \tag{1.1}$$

where the variable u represents the temperature at the position (x, y) in the material. The variable u depends both on time, t, and the spatial position defined by the variables x and y. The quantities c and f are constant parameters in the simplest form of the heat equation but could be functions of the spatial coordinates x and y.

Distributed parameter models are discretised in order to allow conventional simulation tools to be applied. This involves all partial derivatives being expanded and approximated by sets of algebraic equations and differential equations at discrete points to give a set of DAEs that can be handled using standard tools. Techniques commonly used to discretise partial differential equations for simulation include finite element methods, finite difference methods and the method of lines, analytical methods, the integral approximation method, Padé approximation methods, the Ritz method and Galerkin's method. Further details of these techniques may be found in texts dealing specifically with the solution of partial differential equations (see, e.g. [18]). Some simulation environments, such as MapleSim™ [19] include tools that can be used to discretise PDEs and automatically generate components that can be used for simulation. For example, recent developments [20] have provided a method of incorporating PDEs within a Modelica model by using the Functional Mock-up Interface (FMI) [21] of Modelica to import a PDE solver from the HiFlow3 multi-purpose finite element library written in C++ [22]. With further extensions it is believed that this could provide a relatively simple approach which allows re-use of existing software which is known to be efficient and to have been fully verified [20].

In some specific cases it is possible to use analytical techniques to reduce a description based on partial differential equations to a model involving a lumped approximation. Whether or not it is appropriate to approach the model development in this way depends on the application for which the model is being developed. One example of this is the reduction of a distributed parameter model to a lumped representation involving a pure time delay.

A useful set of papers on the modelling, analysis and control of distributed parameter systems may be found in a special issue of the journal *Mathematical and Computer Modelling of Dynamical Systems* published in 2011 with guest editors Kurt Schlacher and Markus Schöberl of the Johannes Kepler University of Linz [23]. The papers in that special issue relate both to theoretical problems concerning the development of distributed parameter models and to a number of applications, including the modelling of flexible structures for sub-sea applications [24].

1.5.2 Discrete-Event and Hybrid Models

In discrete-event models changes of the values of system variables are assumed to occur instantaneously and in a discontinuous fashion at specific instants of time. Such a representation is clearly approximate since real physical variables cannot change instantaneously and the idea behind discrete-event models is to make the model more tractable and to speed-up the simulation significantly. The variables of a discrete-event model change value at specific points in time and these are termed events. Values of variables remain constant between events.

One important example of discrete-event modelling concerns the dynamics of manufacturing systems. Problems in that field can be especially challenging when they relate to the overall dynamics of a network of interacting manufacturing systems and the associated supply chains. In most practical applications of this kind there are a number of well-defined steps associated with fabrication, testing, assembling and packaging. In all kinds of manufacturing the total flow time is influenced by many different factors, such as the processing time at each stage of manufacture, the processing time in testing, the transport time between stages of manufacture and testing and the availability of resources. Flow times in a process like this can also be dependent in a nonlinear fashion on the number of products going through the system. One approach to the development of models for a system of this kind involves the use of "effective" processing times which can be found from observations of a real factory [25].

It should be noted that, although discrete-event models are well-suited to the types of situation that arise in the modelling and simulation of manufacturing systems, other modelling approaches can also be used for such problems. For example, an interesting alternative approach, which is believed to have possible advantages for complex manufacturing situations such as those arising in the semiconductor industry, is based on the use of a model based on partial differential equations. This form of description incorporates both "flow time" and "throughput" [26] and in a model of this kind the flow of products is considered analogous to fluid flow. Similar concepts could clearly be applied to problems in other areas, such as road traffic modelling and control.

As already mentioned, hybrid models bring together continuous system simulation models and discrete-event models. They arise in many different types of application and examples can be found in engineering systems which involve a combination of inherently discrete processes such as those that occur within a digital processor and continuous elements such as sensors and actuators connected to that processor through analogue-to-digital and digital-to-analogue converters.

1.5.3 Inverse Models and Inverse Simulation Methods

Turning the normal process of model development around to find the inputs required to produce specified time histories of chosen variables is a paradigm shift that is increasingly being recognised as offering interesting benefits in some specific areas of application, such as those involving a human operator within a control loop. In an engineering context, the resulting description is termed an inverse model and, in cases where the process does not lead to a mathematical description but is entirely computer-based, the term inverse simulation may be applied. Inverse models and inverse simulations have been used, so far, mainly for problems involving continuous system descriptions based on ordinary differential equations or differential algebraic equations. In principle, the concepts should also apply to discrete-event and hybrid models.

It should be noted that in some fields of application, such as the environmental sciences, "inverse modelling" is a term that is used to describe experimental modelling procedures where a model is fitted to measurements or observations. In this book, following the practice used areas such as mechanical, electrical and aeronautical engineering, such experimental modelling methods are referred to as "system identification and parameter estimation" techniques and "inverse modelling" and "inverse simulation" are terms used to describe the process of finding inputs needed to give a specific form of output time history.

Inverse models are relatively easily developed in the case of systems which behave in a linear fashion. In the case of single-input single-output systems, transfer function descriptions may be inverted simply by interchanging the numerator and denominator polynomials, provided the resulting inverse model is realisable. The issue of realisability arises because models of practical systems often have more poles than zeros in a transfer function description and, since the zeros of the original system model become the poles of the inverse model, the transfer function of the inverse may be found to have fewer poles than zeros. Such a situation means that the inverse transfer function is unrealisable and cannot be implemented as a simulation unless additional factors can be introduced into the denominator of the inverse to make the number of poles equal to or greater than the number of zeros. This can be achieved very readily by adding extra poles into the inverse transfer function in such a way that they have a negligible effect on the overall behaviour of the inverse model but do allow a simulation to be implemented. Such extra poles must lie at points in the left-half of the complex plane and must be far removed from the positions of the poles and zeros of the original model (see, e.g. [27]). Analytical techniques may also be applied for inversion of linear multi-input multi-output models [27]. Inversion techniques for nonlinear models have also been developed using transformations to linear and controllable forms through nonlinear state feedback methods and the use of concepts from differential geometry (see, e.g. [28, 29] Although such analytical methods have been applied with success for some applications involving automatic control and especially aircraft flight control (see, e.g. [30]) relatively little routine use appears to have been made of these highly mathematical techniques. For practical problems involving nonlinear systems simulation-based methods of inversion, which avoid the analytical complexities, are often preferred.

A number of techniques for inverse simulation are based on discretised descriptions of state-variable models with iterative solutions based on gradient information (e.g. [31, 32]) or search-based optimisation methods [33]. Other methods involve continuous system simulation concepts using approaches based on approximate differentiation (see, e.g. [34]) or feedback principles (see e.g. [35, 36]). Inverse simulation methods are discussed in the case study in Chap. 9 relating to the modelling of a system involving two interconnected tanks of liquid.

One interesting fact is that concepts associated with inverse models have recently started to be discussed in connection with physiological systems and with neuromuscular control and voluntary movement in particular. Physiological researchers are suggesting that the human central nervous system may provide combined feedforward and feedback adaptive control (see, e.g. [37, 38]). The feedforward control pathways are believed to be particularly important for executing fast movements as neural delays in feedback pathways would make feedback control ineffective on its own. It has even been suggested that an internal feedback loop is used within the nervous system to implement a form of internal inverse model to provide the feed-forward component of the motor response [39, 40]. This is of special interest in that, as noted in the paragraph above, one of the methods of inverse simulation currently being promoted in the modelling and simulation literature is based upon the use of feedback loops (see, e.g. [35]). It is fascinating to find that a similar feedback-based approach may also have evolved within the human body. This is also relevant for the case study in Chap. 12 relating to modelling of the neuromuscular system.

1.6 Interactions Between Different Types of Simulation Model and Other Software Tools

In some areas of application different types of simulation model may be use within a single project. This can arise, for example, in design and manufacture of an engineering system where design processes may involve the use of continuous system simulation models in conjunction with other discrete-event or hybrid models. Similar issues can arise in projects which have scientific objectives. Whatever the application area, complex simulation projects may involve the use of other types of software for specialised applications such as three-dimensional modelling, visualisation or computer-aided design (CAD). These different activities must be coordinated properly and must function in a fashion that is, as far as possible, completely transparent to the user. It is particularly important to ensure that data transfer between the different software environments is free from errors. This requires careful planning and good software management. One particularly important part of the software management process is to ensure that when design changes are made all the relevant models and software systems are updated together so that every aspect of the project remains in step. Good documentation is an essential requirement in supporting activities of this kind that can involve the use of a variety of different software tools.

1.7 Organisation of the Book

Chapter 2 builds upon the introduction that this first chapter provides and presents an overview of the main concepts of simulation model testing, verification and validation. Issues of model quality, modelling uncertainties and errors are introduced and the iterative nature of the whole modelling process is emphasised. Model evaluation is discussed, both at the sub-model level and in the context of complete system models.

Measures of model quality are discussed in Chap. 3, with emphasis both on graphical methods and deterministic objective measures involving time-domain and frequency-domain data. Statistical measures are also discussed, briefly, together with methods for the efficient visualisation of results of tests.

Chapter 4 is concerned with the use of parameter sensitivity analysis methods in model evaluation processes while Chap. 5 deals with experimental data for model validation. A brief overview of system identification and system parameter estimation techniques is included within that chapter and the importance of issues of identifiability and choice of test input signal for system identification and model validation experiments is emphasised.

Chapter 6 is concerned specifically with issues of model verification, which is the part of the testing procedure for checking that simulation software provides an accurate representation of the mathematical and logical aspects of the model and that the simulation model is based on appropriate algorithms. Various approaches are considered, including the use of formal methods.

Chapter 7 addresses the complex issues of validation or invalidation of models. Techniques considered include simple methods involving comparisons of model predictions and corresponding measurements from the real system, model distortion techniques, barrier certificate methods, system identification and parameter estimation, methods involving sensitivity analysis and inverse simulation. Face validation methods that depend on the knowledge, experience and opinions of people who have expert knowledge of the real system are also discussed. Problems associated with the validation of sub-models and generic models are considered, as are questions of model upgrading and acceptance.

Some vitally important issues of management for the processes of verification, validation, accreditation and application are discussed in Chap. 8. Documentation and cost issues are given particular emphasis.

Chapters 9, 10, 11, and 12 are all devoted to case studies which provide an illustration of the way in which some of the methods of approach described in the earlier chapters can be applied. Two of the case studies are concerned with engineering systems while the others relate to physiological system modelling. All of these case studies relate to areas of research in which the author has worked.

Chapter 9 is based on an application involving a laboratory scale process system involving two coupled tanks of liquid. A nonlinear model of the system is presented and experiments are designed to allow values of some parameters of the system to be estimated. Verification and validation issues are then discussed for this system application and results are presented for a number of different methods. The case study concludes with discussion of possible improvements to the basic simulation model.

Chapter 10 gives an account of model validation issues in the context of helicopter flight simulation. This is a field in which simulation has an important role, both in vehicle design and in the development of flight simulators for use in pilot training and in research. System identification and parameter estimation techniques have proved particularly important in this type of application. Issues of model structure and test input design are emphasised. In helicopter applications there are important constraints on the form of test input that can be applied, both in terms of the magnitude of the test signal and the maximum duration of the flight experiment and practical issues of this kind are considered carefully in this case study.

The third case study, in Chap. 11, is concerned with modeling and simulation of the gas exchanging properties of the human lungs. The cardio-respiratory system is a physiological system which has attracted much attention in terms of modelling and simulation. Most models that have been developed are compartmental in form and have been derived from physical, anatomical and physiological knowledge. This case study provides an interesting illustration of the importance of identifiability analysis and also of test signal design where, as in the helicopter case in Chap. 10, there are significant practical constraints in terms of the duration of the experiment and the magnitude of the applied test input. Applications considered include the use of simulation models for the non-invasive estimation of physiological quantities such as lung volume and cardiac output.

Chapter 12 is another physiological case study and deals with simulation models of neuromuscular systems. A simple empirical model of muscle is presented and issues relating to the problems of model validation from limited experimental data are discussed. Issues arising in the validation of models of other elements of the neuromuscular system are then considered. The chapter concludes with an account of the contribution that modeling and simulation techniques are making in the testing of hypotheses concerning the control of movement and regulation of posture, especially in terms of the role of some specific elements within the peripheral and central parts of the nervous system.

The book concludes, in Chap. 13, with a brief discussion which attempts to bring together some of the main issues relating to the principles and practice of simulation model validation in the light of the case studies chapters. It then goes on to provide an assessment of some of the current strategic issues in the field of modelling and simulation, together with some discussion of current trends in terms of model testing and validation. Research and development opportunities are also discussed, together with some important issues in terms of education.

References

1. Foss BA, Lohmann B, Marquardt W (1998) A field study of industrial modelling process. Model Identif Control 19(3):153–174
2. Famme JB, Gallagher C, Raitch T (2009) Performance based design for fleet affordability. Nav Eng J 12(4):117–132. doi:10.1111/j.1559-3581.2009.00233.x
3. Ericsen T (2004) Guidance for the development of ship-smart system design (S3D), ONR BAA #04.001, Advanced Integrated Power Systems, Boston, July 2004
4. Hemez FM (2004) The myth of science-based predictive modelling. In: Proceedings foundations '04 workshop for Verification, Validation and Accreditation (VV&A) in the 21st century, Arizona State University, Tempe, Arizona, October 13–15 2004. Report LA-UR-04-6829, Los Alamos National Laboratory, USA
5. Cloud DJ, Rainey LB (eds) (1998) Applied modeling and simulation: an integrated approach to development and operation. McGraw-Hill, New York
6. Bruenese APJ, Top JJ, Broenik JF et al (1998) Libraries of reusable models: theory and application. Simulation 71(1):7–22

7. Fritzson P (2004) Principles of object-oriented modeling and simulation with Modelica 2.1. IEEE Press, Piscataway
8. Matlab®/Simulink® modelling and simulation software. Mathworks Inc. http://www.mathworks.com/products/. Accessed 5 June 2015
9. Simscape™ software. Mathworks Inc. http://www.mathworks.com/products/. Accessed 5 June 2015
10. VTB modelling and simulation package. The University of South Carolina. http://vtb.engr.sc.edu/vtbwebsite/. Accessed 5 June 2015
11. McKay W, Monti A, Danti E et al (2002) A co-simulation approach for ACSL-based models. Paper presented at Huntsville simulation conference 2002, Huntsville, 9–10 October 2002. http://www.scs.org/confernc/hsc/hsc02/hsc/papers/hsc035.pdf. Accessed 5 June 2015
12. Christen E, Bakalar K (1999) VVHDL-AMS – a hardware description language for analog and mixed-signal applications. IEEE Trans Circuits Syst II: Analog Digit Signal Process 46 (10):1263–1272
13. Cohen A (1996) Generic simulation models of communication systems. In: Proceedings 29th annual simulation symposium, New Orleans, LA, USA, 8–11 April 1996. IEEE, USA, pp 81–88. doi:10.1109/SIMSYM.1996.492155
14. Andreasson J (2007) On generic road vehicle motion modelling and control. PhD thesis, Royal Institute of Technology, KTH, Stockholm. http://kth.diva-portal.org/smash/get/diva2:11627/FULLTEXT01. Accessed 5 June 2015
15. Smith MI, Murray-Smith DJ, Hickman D (2007) Mathematical and computer modeling of electro-optic systems using a generic modeling approach. J Def Model Simul 4(1):3–16
16. Stein K, Walther SM (2013) A generic simulation model for planning critical care resource requirements. Anaesthesia 68(11):1148–1155. doi:10.1111/amae.12408
17. European Space Agency, Modelling and Simulation. Generic Project Test Bed. http://www.esa.int/TEC/Modelling_and_simulation/SEMTRH8LURE_0.html. Accessed 5 June 2015
18. Pinchover Y, Rubenstein J (2005) An introduction to partial differential equations. Cambridge University Press, New York
19. Maple™ software; Maplesim™ modelling and simulation software: Maplesoft, a division of Waterloo Maple Inc. www.maplesoft.com/products/maplesim. Accessed 5 June 2015
20. Staväker K, Ronnå S, Wlotzka M et al (2014) PDE modelling with Modelica via FMI import of HiFlow3 C++ components. Simul Notes Eur 24(1):11–20. doi:10.11128/sne.24.tn.102223
21. Modelica Association (2012) Functional mock-up interface for model exchange and co-simulation, v 2.0 beta 4 edition
22. The HiFlow3 multi-purpose finite element software, Engineering Mathematics and Computing Lab (EMCL), Heidelberg University. http://www.hiflow3.org. Accessed 5 June 2015
23. Schlacher K, Schöberl M (eds) (2011) Special issue: Modelling, analysis and control of distributed parameter systems. Math Comput Simul Dyn Syst 17(1):1–121
24. Forteleza E, Creff Y, Levine J (2011) Active control of a dynamically positioned vessel for installation of subsea structures. Math Comput Simul Dyn Syst 17(1):71–84
25. Jacobs JH, Teman LFP, van Campen EJJ et al (2003) Characterization of the operational time variability using effective processing of times. IEEE Trans Semicond Manuf 16(3):511–520
26. Lefeber E, van den Berg RA, Roods JE (2004) Modeling, validation and control of manufacturing systems. In: Proceedings of 2004 American control conference, Boston, June 30–July 2, 2004, pp 4583–45880
27. Buchholz JJ, von Grünhagen W (2004) Inversion impossible? Technical report, University of Applied Sciences Bremen, Germany, September 2004
28. Isidori I (1989) Nonlinear control systems: an introduction, 2nd edn. Springer, Berlin
29. Hunt LR, Meyer G (1997) Stable inversion for nonlinear systems. Automatica 33(8):1549–1554
30. Zou Q, Devasia S (2007) Preview-based inversion of nonlinear nonminimum-phase systems: VTOL example. Automatica 43(1):117–127

31. Hess RA, Gao C, Wang SH (1991) A generalized technique for inverse simulation applied to aircraft maneuvers. J Guid Control Dyn 14:920–926
32. Thomson D, Bradley R (2006) Inverse simulation as a tool for flight dynamics research – principles and applications. Prog Aerosp Sci 42:174–210
33. Lu L, Murray-Smith DJ, Thomson DG (2008) Issues of numerical accuracy and stability in inverse simulation. Simul Model Pract Theory 16:1350–1364
34. Murray-Smith DJ (2013) An approximate differentiation method of inverse simulation based on a continuous system simulation approach. Simul Notes Eur 23(4):105–110
35. Murray-Smith DJ (2011) Feedback methods for inverse simulation of dynamic models for engineering systems. Math Comput Model Dyn Syst 17:515–541
36. Tagawa Y, Tu JY, Stoten DP (2012) Inverse dynamics compensation via 'simulation of feedback control systems'. Proc Inst Mech Eng Part I: J Syst Control 225:137–153
37. Donaldson R, Jones RD, Andreae JH et al (2002) Simulating closed- and open-loop voluntary movement: a nonlinear control systems approach. IEEE Trans Biomed Eng 49(11):1242–1252
38. Miall PC, Wolpart DM (1996) Forward models for physiological motor control. Neural Netw 9:1265–1278
39. Davidson PR, Jones RD, Sirisena HR et al (2000) Detection of adaptive inverse models in the human motor system. Hum Mov Sci 19:761–795
40. Yavari F, Towhidkhah F, Amadi-Pajouh MA (2013) Are fast/slow process in motor adaptation and forward/inverse internal model two sides of the same coin? Med Hypotheses 81(4):542–600

Chapter 2
Concepts of Simulation Model Testing, Verification and Validation

The key question in modelling is the quality needed for the application in hand and the adequacy of the chosen model for that specific application. Errors must be kept within specified limits for particular parts of the operating envelope. Testing, verification and validation may be viewed as processes that allow boundaries to be defined in terms of model performance. This chapter addresses issues of model quality and describes, in a general way, the iterative processes of model development, testing and acceptance. It thus provides a link between the broadly-based discussion of simulation model development processes in the first chapter and more detailed consideration of measures of model quality in Chap. 3 and methods of testing that are described in later chapters.

2.1 Model Quality, Uncertainties and Errors

It should be noted that the word "validation" is often used in a rather misleading way, implying that it is possible to establish, once and for all, that a given model is "correct". Anyone with modelling experience will know that what we are really interested in when we are assessing model quality is more related to a process of invalidation. Although it is relatively straightforward to prove that a model is invalid for some set of conditions it is impossible to prove that it is "valid". At best we may be able to demonstrate that a given model appears to be an acceptable representation of a real system for the specific purposes of the intended application. In that very special sense the model may be viewed as being a valid representation but it must always be remembered that new evidence may be found at any time that may invalidate the model. In engineering design and development applications the use of models that are inadequate for an intended application can lead to expensive redesign at late stages in the development of a system. In science the use of inappropriate models is clearly counter-productive, may take investigations in an

D.J. Murray-Smith, *Testing and Validation of Computer Simulation Models*, Simulation Foundations, Methods and Applications,
DOI 10.1007/978-3-319-15099-4_2

inappropriate direction and generally impedes progress. The more complex the model being considered, the more likely it is that problems of model inadequacy will arise.

Model building is an iterative procedure and many, including Sargent (e.g., [1]), Ören [2], Balci (e.g., [3–5]), and Brade (e.g., [6, 7]) have pointed out that model testing and evaluation are inseparable from the other processes of model building. These evaluation processes are undertaken to quantify confidence and to build credibility in a numerical model for the purposes of making predictions. Prediction can be defined as the use of a model to predict the state of a physical system under conditions for which the fidelity of the model has not been fully assessed [8]. If necessary, the predictive accuracy can be improved through additional experiments, information and experience [9]. This is emphasised in the structure of Fig. 2.1 where feedback pathways from the outcome of the model testing stage lead back to blocks dealing with the selection of the model structure and the parameter values. Confidence in a model should increase steadily during its development, if appropriate methods are applied and if correct methods of testing are used. What one is attempting to do is to define a neighbourhood of validity within which a model produces results that are consistent with the behaviour of the corresponding real system. Within that neighbourhood simulation may be a substitute for testing of the real system, in the context of the intended application [10].

Simple models are often used for examining "what if" situations in the initial stages of an investigation and, in an engineering context, for early-stage design trade-off studies. Error bounds on model predictions at the first stages of model development are usually large and quantitative validation procedures can seldom be applied at that point. Model quality and fitness-for-purpose can only be assessed on the basis of experience and through comparisons made with earlier models of similar systems, since data from the real system may be very limited. More refined models may be used at a later stage, especially when more data become available for model testing. Once testing and experimentation on the real system has started, more and more quantitative information flows from the real system to the model. In engineering, bi-directional information transfer is a characteristic of the later stages of any model-based development, with model updates being applied as more information about the real system becomes available. In scientific applications bi-directional information flow is also important, with models being refined and used for the design of better experiments to generate additional data from the real system.

The main reasons for uncertainties and errors in models are inappropriate assumptions, errors in parameter values and other aspects of the model, errors in numerical solutions and errors in experimental procedures and measurements. Although the adoption of a strategy in which different parts of the process of model development are clearly separated is always useful, and categorising simulation model errors is important, model uncertainties are inevitable since our understanding of the real system is never complete and our measurements and calculations are limited in accuracy.

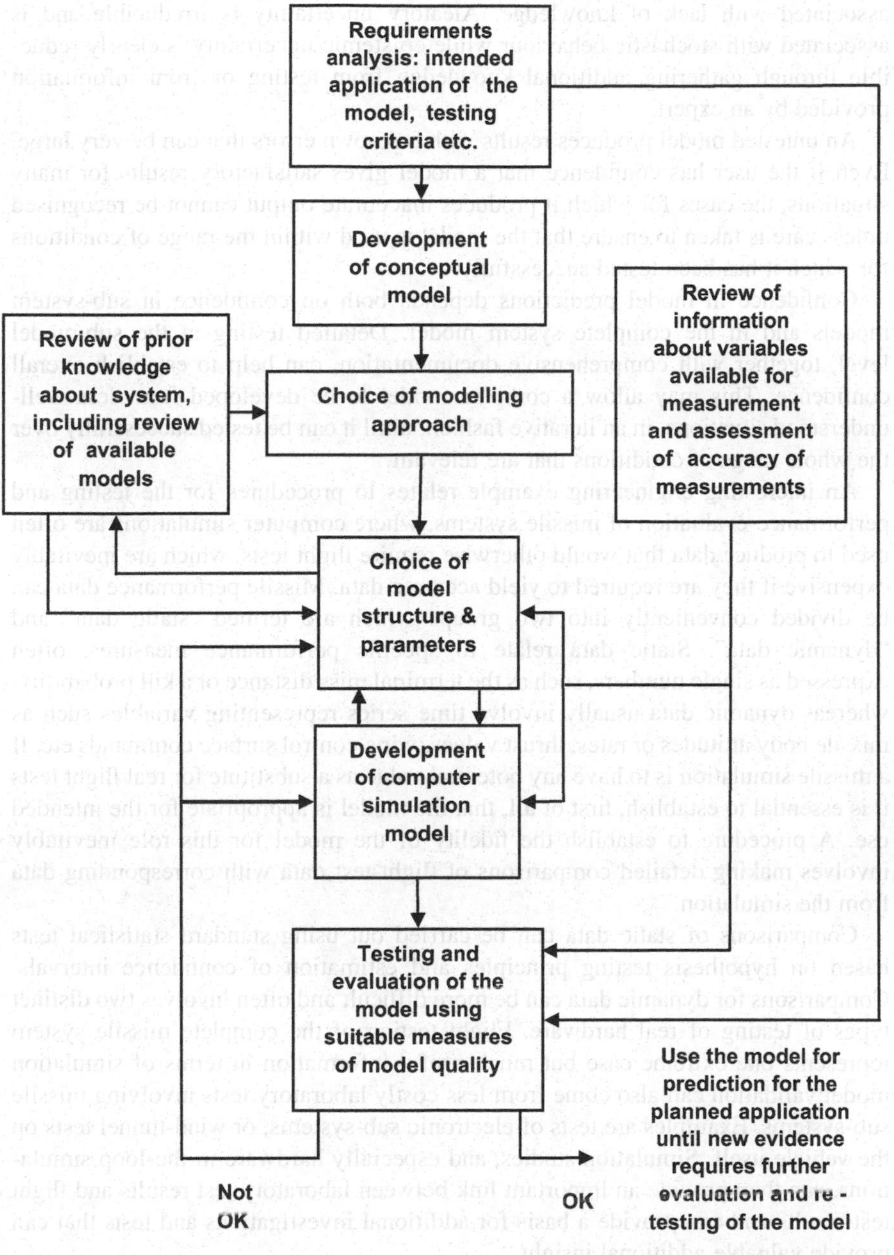

Fig. 2.1 Block diagram showing the initial top-level representation of the modelling process

Oberkampf [11] has divided uncertainties into two broad categories. The first of these he terms "aleatory" uncertainties, which he defines as the inherent variation associated with the physical system and its environment. The second category involves "epistemic" or "subjective" uncertainty which he defines as being associated with lack of knowledge. Aleatory uncertainty is irreducible and is associated with stochastic behaviour while epistemic uncertainty is clearly reducible through gathering additional knowledge from testing or from information provided by an expert.

An untested model produces results with unknown errors that can be very large. Even if the user has confidence that a model gives satisfactory results for many situations, the cases for which it produces inaccurate output cannot be recognised unless care is taken to ensure that the model is used within the range of conditions for which it has been tested successfully.

Confidence in model predictions depends both on confidence in sub-system models and in the complete system model. Detailed testing at the sub-model level, together with comprehensive documentation, can help to establish overall confidence. This may allow a complex model to be developed from less well-understood situations, in an iterative fashion, until it can be tested successfully over the whole range of conditions that are relevant.

An interesting engineering example relates to procedures for the testing and performance evaluation of missile systems, where computer simulations are often used to produce data that would otherwise require flight tests, which are inevitably expensive if they are required to yield accurate data. Missile performance data can be divided conveniently into two groups which are termed "static data" and "dynamic data". Static data relate to specific performance measures, often expressed as single numbers, such as the terminal miss distance or a kill probability, whereas dynamic data usually involve time series representing variables such as missile body attitudes or rates, thrust values, wing control surface commands etc. If a missile simulation is to have any potential value as a substitute for real flight tests it is essential to establish, first of all, that the model is appropriate for the intended use. A procedure to establish the fidelity of the model for this role inevitably involves making detailed comparisons of flight-test data with corresponding data from the simulation.

Comparisons of static data can be carried out using standard statistical tests based on hypothesis testing principles and estimation of confidence intervals. Comparisons for dynamic data can be more difficult and often involves two distinct types of testing of real hardware. Flight testing of the complete missile system represents one extreme case but much useful information in terms of simulation model validation can also come from less costly laboratory tests involving missile sub-systems. Examples are tests of electronic sub-systems, or wind-tunnel tests on the vehicle itself. Simulation studies, and especially hardware-in-the-loop simulations, can then provide an important link between laboratory test results and flight test results and can provide a basis for additional investigations and tests that can provide valuable additional insight.

2.2 The Iterative Processes of Model Development, Testing, Improvement and Acceptance

Once the evaluation stage is reached in the model development process, the developer is faced with issues that relate to the adequacy or otherwise of the model for the intended application. Although quantitative measures can be very helpful in the decision making process, there are often more subjective issues that also need to be taken into account and the intended end-user of the model may have an important role in this. For example process plant operators have a wealth of experience with real process plant and may be able to suggest testing situations that, from their practical perspective, may be of critical importance in making an assessment of a plant model. Similarly, a medical specialist may be able to contribute to the assessment of a physiological model by suggesting that the behaviour of the model should be looked at for test conditions that correspond to specific clinical conditions or tests. Provided these clinical situations are relevant in terms of the intended application of the model, investigations of that kind can be very revealing and can be very helpful in identifying model deficiencies.

If a specific model, for whatever reason, is found to be inadequate in terms of the use for which it has been developed there is an obvious need for upgrading of the model. As already suggested in Chap. 1 this can involve the use of measured data sets from the real system, when these are available, for the purposes of system identification and parameter estimation. This may then allow one to assess the possible benefits to be obtained from a change of structure or model parameters. More fundamental changes in terms of the assumptions or approximations inherent in the model may also be required. Tools for sensitivity analysis can be very important in that context and are discussed in Chap. 4.

In some situations the investigator may be faced with a choice between using a relatively simple model and one that appears much more complex but may have a structure that relates more closely to reality. For the immediate application the two models may show little difference in terms of the results of tests of model adequacy and one might be tempted to set aside entirely the more complex model and concentrate efforts on the simpler representation. This could be found at a later stage to be inappropriate as one seldom has all the information available in the early stages of a project about all the possible ways in which a model may have to be used. In other words, one should be aware of the possibility of changes being needed in the intended application of the model. Keeping alive a range of models of different complexity may be very helpful and different versions of a model of a given system may provide valuable insight that would be hard (and perhaps costly) to obtain in other ways.

The issue of model complexity is often discussed in the context of the philosophical ideas associated with Occam's "razor". William of Occam was a Franciscan friar and logician who lived in the fourteenth century and was responsible, among many other things, for a statement *Frustra fit per plura quod potest fieri per pauciora*. When translated from the Latin this means, roughly, that when there are two competing theories which make the same predictions the simpler theory is always the better. This proved to be a very powerful idea in early scientific advances in fields such as astronomy and its value is often still emphasised in the context of modelling. However, it has been pointed out recently that, in fields such as biology, the indiscriminate use of Occam's razor in the context of system modelling has led to serious mistakes. The razor should be used to cut out elements of theory that cannot be observed or measured experimentally to leave a simpler and more heuristic model. However, in most areas of biology models end up being far more complicated than expected at the outset of a project and model simplicity is seldom an outcome encountered in practice since evolutionary processes do not necessarily lead to simple solutions. Indeed in some areas, such as pulmonary gas exchange modelling, it has been suggested that Occam's razor has been used to justify cutting out elements, for which information is available, for reasons of computational convenience rather than on an objective basis linked to questions of model quality [12].

Figure 2.1 shows a block diagram of the complete modelling process, starting from the intended application and the associated requirements analysis and the prior knowledge available about the real system under consideration. Information is included about any earlier models that may be available, as well as information about variables that can be measured in the real system. The review of the intended application must include consideration of those who are expected to make use of the model (the user "community"). These three aspects of the model are of fundamental importance. Once they have been considered the process of model formulation can be started. Both the intended application and prior knowledge about the real system can have a significant influence on the form of model that is developed and both these aspects affect the choice of modelling approach. Once an approach has been selected, the initial choice of model structure and parameter values needs to be made.

It is clear that the block diagram shown in Fig. 2.1 contains no details and is a highly simplified representation. What it is particularly important about it is the fact that it includes a block that represents the testing of the simulation model and also that there are feedback pathways from the testing procedures to the blocks concerned with the development of the mathematical and logical model of the system and the development of the simulation model. This emphasises the fact that the model development process is essentially iterative and that it is influenced not only by the *a priori* information and the intended usage of the model but also by the available information about the real system including measurements.

From this stage in the process, questions about model testing and model adequacy start to become important and appropriate testing methods need to be defined and the detailed processes of simulation model development and procedures for model testing need to be taken forward together. These issues are discussed in more detail in later chapters.

When a simulation model is being developed for an application involving a contract from a government agency there is often an official certification process known as "accreditation" that has to be followed. This is simply a recognition that a model and the associated computer simulation (and the data required for it) are acceptable for use in the context of the intended application and have been subjected to an approved verification and validation procedure (see, e.g. [13]). In most cases the government sponsoring organisation does not specify exactly how each stage of the testing process should be carried out and instead provides broad guidance.

A degree of confidence needs to be established in the model so that its results can be recognised as being reasonable for the objective for which it has been developed. Circumstances in which the model ceases to be useful must also be established. The inherent errors and uncertainties in the simulation model must be understood and, ideally, minimised. However, it is also important to recognise that experimental data used in the development and testing of the model will also have inherent errors.

The development and testing of models generally starts at the component level and moves gradually towards testing of a complete system model. Ideally one wants to move from a relatively simple and well-understood situation towards cases involving more uncertainty. An interesting case study is provided by an investigation at Argonne National Laboratory involving the validation of simulation models of hybrid electric vehicles [14]. The approach used was based on splitting a complex integrated system into a set of sub-systems which could be modelled separately, with each sub-system model being subjected to a rigorous testing process.

The eventual goal in the testing of simulation models is usually to maximise the confidence in the predictive capabilities of the model. Errors can arise in at least two different ways, since deriving a computer simulation can give rise to errors in addition to those that can arise in developing the model itself. The word "verification" is commonly used to describe the process of establishing that a simulation and its associated data are consistent with the underlying mathematical model, while the word "validation" describes the process of determining the degree to which a model and the associated data set are an accurate representation of the real world from the perspective of the intended model application. These conventions are consistent with recommendations established as early as 1979 by the SCS Technical Committee on Model Credibility [15]. Although the SCS recommendations have been widely used for many years, the words "verification" and "validation" are still applied very loosely in some fields. In an attempt to reduce uncertainties in the terminology the present author suggested associating the word "internal" with the word "verification" and "external" with the word "validation". This was intended to make it clear that "internal verification" was concerned with the self-contained process of checking the consistency of the computer simulation model with the underlying mathematical and logical description and "external validation" involved checking consistency of the model with the external real-world system that it represented [16]. This was seen by some as being helpful at the time and it is interesting to see that the health care modelling community in recent recommendations have also use the term "external validation" with a similar meaning (see, e.g. [17]). Unfortunately, it appears that the same community has adopted the term "internal validation" to describe what is more broadly accepted as a verification process. Problems of terminology continue to plague us!

From the time of the SCS recommendations on model credibility [15] many different diagrams have been employed to illustrate the processes of simulation model verification and validation, but it should be noted that some of the simpler and more concise diagrams of this kind do not show clearly that model validation is a part of the much larger and essentially iterative process of model development and many also do not show explicitly the pathways for upgrading a model. Validation must be recognised as an on-going activity which is never complete and can continue throughout the whole life cycle of the system being modelled. Many simple diagrams also fail to emphasise properly a number of important aspects of the model testing process, including the parallel nature of experimentation and model development and the presence of uncertainties in both the simulation and experimental results.

In recent years there has been a resurgence of interest in issues associated with model testing, verification and validation. In an engineering context this has been due, partly, to initiatives such as the US Department of Energy Accelerated Strategic Computing Initiative (ASCI) which has involved significant investment in research and development relating directly to verification and validation methods and tools [18]. Another important development has been the establishment, about 2001, of the American Society of Mechanical Engineers (ASME) Committee on Verification and Validation in Computational Modeling and Simulation. This Committee involves four groups dealing with more specific topics in application areas such as nuclear system thermo-fluids, medical devices and solid mechanics. Some of these, such as the medical devices group have additional sub-groups dealing with specialist topics such as modelling of heart valves and stents [19]. This ASME Committee and its sub-groups have been responsible for the publication of a number of standards and related reports on verification and validation issues and have a number of additional standards documents in draft form at the time of writing [20].

Other organisations that have made important recommendations or guidelines relating specifically to simulation model validation and verification (V&V) include the AIAA Computational Fluid Dynamics V&V Committee [8] and the Modeling and Simulation Coordination Office (M&SCO) (which was, until 2007, called the Defense Modeling and Simulation Office (DMSO)) of the US Department of Defense [21]. Various research groups within the US Department of Energy laboratories, such as Sandia, Los Alamos and Lawrence Livermore, have also published the results of recent research on model quality and validation (see, e.g. [9]).

It is important to note that model testing and the processes of verification and validation are activities which are distinct from software testing. Model verification and validation processes are used when a predictive model is the required end product, while software verification and validation processes are used when a computer program or other software is the end result. A model always includes more than the code that forms the associated computer programs. Software testing for that code is, of course, still essential and the processes of verification and validation do not replace that more basic testing process. Simulation model verification processes should include checks to establish whether earlier software testing has been carried out effectively. For example, it would be expected that any coding errors picked up in the verification process for the simulation model would be associated with simulation-related coding issues rather than basic programming mistakes.

In the more general software testing field there has been a growth of interest in recent years in testing techniques which are known as "formal methods". In application areas involving systems that are in some way "safety critical", a failure of system software could have disastrous consequences in terms of damage and possible loss of life. Many examples of safety critical systems can be found in embedded systems, such as are found now in road vehicles, in aircraft systems, in medical equipment, in nuclear power generation and in industrial process applications. For such applications it has been suggested that formal methods could provide important benefits and much research has been under way in this field for some considerable time.

It has been suggested that this approach may also be of potential value for the purposes of model verification and validation [22]. Although there are some applications, involving relatively simple models, where ideas from formal methods have shown promising results when applied to modelling problems, especially in terms of analysis of model requirements and the associated conceptual models, the approach is not yet one that has been widely applied. These methods, which attempt to prove in a mathematical sense that a system meets its specification, are expensive to apply, both computationally and in terms of the specialist skills and human effort required. Indeed, in the context of model verification and validation, doubts have been raised about the potential of formal methods for anything beyond requirements analysis and the assessment of software within a model (i.e. the verification aspect of model testing) [23]. Although extremely effective for small problems, many formal techniques are regarded as unusable for other than trivial applications. The topic of formal methods is discussed again, briefly, in Chaps. 6 and 7, dealing with verification and validation methods.

Another safety-critical type of application relates to the use of models in safety assessments of nuclear waste repositories. The use of modelling and simulation techniques for assessing the performance of potential sites for storage of nuclear waste are linked closely to repository design issues and to characterisation of the site in terms of its geology. The development of a conceptual model is, as always, an important first step and, for most systems involving a combination of engineering solutions and natural storage, the development of a single model is unlikely. A more probable scenario would involve several conceptual models that go some way towards meeting all the known requirements and satisfying all of the known constraints. The complication is that, while engineering systems can be designed to satisfy given sets of requirements, natural geological structures must be explored and characterised and this process can never lead to a model that is in any sense perfect or exact. Any kind of site testing is also likely to be limited in scope because of concerns that testing could be "destructive" and could reduce the value of the site as a potential storage facility. In an application of this kind a hierarchy of models is needed. Some of these are detailed mechanistic process models describing specific sub-systems that are believed to be relevant in gaining an appropriate level of understanding of the problems posed within a site. These may be termed "Level 1" models. At Level 2 we have models which may couple together a number of the detailed Level 1 models to help to improve understanding of interfaces between processes. At the third (and top) level the second-level models are linked together (possibly after some simplification, where judged appropriate) to provide a total system model. These can then, eventually, provide a basis for probabilistic investigations of repository performance after careful consideration of fitness for purpose at each model level. Uncertainties inherent in model structures and parameter values mean that the result of field observations, combined in some cases with laboratory experiments using test rigs that represent specific detailed features of Level 1 models, must usually be the closest that one can come to providing evidence to support modelling assumptions. Future problems within a repository system may be as a result of human actions or may involve natural events. Scenarios therefore have to be defined using the repository models to assess the probability of the release and transport of potentially harmful material. Tests of compliance with regulatory requirements are then judged in terms of probabilistic measures and "reasonable assurance" that the facility will be safe. Establishing credibility is the over-riding requirement, with a need to demonstrate that the scientific and engineering basis is adequate and that each model is fit for purpose in terms of accuracy [24].

The assessment of models used for investigating potential sites for repositories involve tests of compliance with regulatory criteria that apply within the relevant country. Where information is available, these tests of compliance appear to be written in terms of "reasonable assurance" and recognise that absolute assurance of compliance with regulatory criteria is impossible.

2.3 The General Principles of Model Evaluation

Use of the word "validation" may give a false impression of model capabilities and the term "model evaluation" is often more appropriate. Theories can be proved to be wrong but cannot ever be proved to be right and there is always a risk of false confidence in model-based predictions if the model involved appears to have been subjected to some form of "validation". Models that can provide accurate predictions of the performance of the corresponding real system for some restricted circumstances do not mean that the model can give good predictions in all cases.

Errors in modeling can arise from many sources. These include incorrect modeling assumptions, errors in *a priori* information such as parameter values, errors in numerical solutions of the model equations and also errors in the experiments used as the basis for model testing. Separating out the many different possible sources of error is a major challenge. Essentially we have to keep on conducting tests and evaluations until we are sufficiently confident that the model is acceptable for the application being considered.

The requirement that the validation process should establish the predictive accuracy of the model underscores the key role of uncertainty quantification in this process. Uncertainties exist in the outputs of computational simulations due to inherent or subjective uncertainties in the model and measurements made to validate these simulation outcomes also contain errors and uncertainties [9]. Although the experimental results are used as a form of reference for comparison with the simulation outputs, the V&V processes do not make assumptions about correctness. The goal is to quantify the uncertainties in both simulation and experimental results so that model fidelity requirements can be assessed and the predictive accuracy quantified.

A model that has been assessed successfully for a specific application may be judged "fit for purpose" but the inevitable presence of "unknown unknowns" means that there can never be a simple conclusion to an evaluation process of this kind. Tests can only deal with a small number of cases and general statements about validity must be avoided. The more complex the model the greater the difficulties associated with quality assessment: measures of model performance become harder to define and visualisation becomes more difficult.

It should be noted that traditional experiments used to gain an initial understanding of the physical phenomena involved in a given model may differ significantly from experiments designed specifically for model validation. In the latter case the experiment is designed (or should be) to provide quantitative estimates of the ability of the model to simulate the real system and must capture the essential features of that system. In terms of dynamic systems this means that the frequency domain characteristics of test inputs used for validation experiments must be chosen using whatever information is already available about the characteristics of the real system and the intended application of the model. This produces an obvious difficulty since it would appear that, in order to design an optimal experiment for validation purposes, an accurate model of the system is needed. In practice we do not have such a model at this stage and we must make use of the best model available for the purposes of experiment design, as discussed in Chap. 5. Some of the chapters dealing with case studies provide concrete examples of the importance of experimental design for validation.

If evaluation is applied appropriately at each stage of the development of a simulation model, confidence in that model should increase steadily during the development process. As mentioned earlier, we should always attempt to start from a well-understood case, even if much simplified; then move incrementally to testing for more uncertain situations that relate directly to the application.

Once a model has been accepted for the intended application it can be used to make predictions for cases that are slightly different from those for which it has been validated. This is a form of "generalisation", but in order to assess the value of such a prediction one must know how close the set of operating conditions for the prediction is to conditions that applied for cases considered in the validation process [11].

One may also wish to know whether or not the model contains any features that are inconsistent with generally accepted principles of physics, biology or chemistry of the system or process being modelled. This may be termed "physical" or "theoretical" validity and is clearly very important in some circumstances, such as when a model is being used in teaching. In other cases where a model need only be a correct representation of a system in terms of its behaviour for a chosen set of input and output variables we are dealing with "functional" validity. In this situation the primary aim of any evaluation process is to ensure that the model mimics the input-output behaviour of the real system to some acceptable level of accuracy. Assessment of theoretical validity requires critical examination of the structure of the model and is normally a part of a broader evaluation of the model that would include predictive validation tests.

2.4 Verification and Validation of Sub-models

The concepts of verification and validation discussed in earlier sections apply both to complete models of complex systems and to sub-models. In terms of the application, the interest is always in a complete system model which includes all the relevant features of the real system. However, most practical systems include a large number of sub-systems and components, many of which are likely to take the form of hardware but some may well be software sub-systems. Each of these sub-systems must be modelled and each of those sub-models should be subjected to verification and validation processes.

An important issue that arises with systems involving sub-models is that each dynamic sub-model forming a component part of a larger system model must be subject to a process of "initialisation", where reasonable initial values are chosen or calculated that are consistent with the inputs and outputs for the complete system for the chosen operating conditions. This is an important step prior to validation and is normally attempted after verification processes have been completed. Initialisation requires thorough and complete understanding both of the conceptual model and of the mathematical model.

There is an unfortunate tendency to view the validation process as one that only involves test data from the complete system and this leads to difficulties for a number of reasons. One obvious reason is that, in engineering design applications, models of sub-systems may exist long before the complete system model is available. The model validation process can therefore begin, in a natural way, for sub-models initially and move towards validation of the complete system model as the system itself is developed. In physiological applications conducting experiments on sub-systems is, again, a natural way to develop an understanding of a complex biological system.

Sensitivity analysis can be useful in establishing which components and sub-models are of greatest importance within a hierarchical representation such as this but this process must be approached with caution as, inevitably, there are still many modelling uncertainties at this stage and any kind of sensitivity analysis is model dependent. Components and sub-models that are found to show high sensitivities clearly require more attention in terms of validation effort. Review by subject experts can also contribute to establishing the critical elements that require most careful consideration.

In the case of validation tests on a complete system model it can be hard to establish where discrepancies arise if no data are available for sub-model validation. Even if there is acceptable agreement between model predictions and experimental results this is not necessarily conclusive if the only tests are from the complete system and the complete system model. Error cancellation within sub-models could be causing a false impression of agreement. The best approach is to devise a series of tests and experiments that allow confidence to be built up about the quality of the sub-models and of the complete system model. It has been suggested (e.g. [9]) that it should be possible to construct a hierarchy of validation assessments with the top tier representing the complete system and lower tiers the sub-systems and components. Testing should be carried out from the component level upwards, where the sub-models and component models are, in most cases, developed using physical laws and principles, coupled in some cases to the application of system identification and parameter estimation techniques. Data collected at the different levels of this hierarchical structure may come from separate tests carried out on the sub-systems in isolation. If this is the case it is very important that conditions in tests on a sub-system, such as loading or temperature, should be representative of conditions that will apply for the that sub-system in the eventual application. Hardware-in-the-loop real-time simulation methods provide one way of ensuring that the necessary conditions are applied. Issues of this kind are explored in considerable detail in the paper by Pasquier, Duoba and Rousseau of the Argonne National Laboratory (ANL) in the USA [14]. The work described in that paper relates to the modelling of hybrid electric vehicles and discusses how ANL engineers have developed an inter-related set of tools and facilities to analyse, develop and validate components within the highly integrated power-train systems in vehicles of this kind.

2.5 Discussion

In the definition of "validation" given in Sect. 2.2 (where it is stated that this describes the process of determining the degree to which a model and the associated data set are an accurate representation of the real world, from the perspective of the intended model application), the words "process of determining" emphasise the fact that validation is an activity which only reaches a conclusion when agreement between the reference data (often experimental data) and the model output is judged to be acceptable. Also, the words "degree to which" reinforce the fact that some uncertainty still exists, even when a positive conclusion has been reached about the fidelity of the model. The use of the words "intended model application" in the definition further emphasises the fact that the model is limited to a specific application and use of the model for a different application would require further validation assessment. Whatever the intended application, predictions made for conditions close to conditions considered during the validation process will have higher confidence than predictions made for conditions further away from conditions considered during validation.

Any adjustment of model structure or parameters as a result of information obtained during the validation process is not part of the validation procedure and forms part of the model development process and is better referred to as model "calibration". Further validation is necessary after each change of the model and this again emphasises the fact that model development is a process which involves repeated cycles of calibration and testing.

References

1. Sargent RG (1979) Validation of simulation models. In: Highland HJ (ed) Proceedings of the 1979 Winter Simulation conference, vol 2. IEEE Press, Piscataway, pp 497–503
2. Ören TI (1981) Concepts and criteria to assess acceptability of simulation studies: a frame of reference. In: Adam N (ed) Simulation Modeling and Statistical Computing. Communications of the ACM 24(4):180–188
3. Balci O (1997) Principles of simulation model validation. Trans Soc Comput Simul Int 14(1):3–12
4. Balci O (1997) Verification, validation and accreditation of simulation models. In: Andradóttir S, Healy KJ, Withers DH et al (eds) Proceedings of the 1997 Winter Simulation conference. IEEE Computer Society, Washington, DC, pp 135–147
5. Balci O (2004) Quality assessment, verification and validation of modeling and simulation applications. In: Ingailis RG, Rossetti MD, Smith JS et al (eds) Proceedings of the 2004 Winter Simulation conference. IEEE Computer Society, Washington, DC, pp 122–129
6. Brade D (2000) Enhancing modeling and simulation accreditation by structuring verification and validation results. In: Joines JA, Barton R, Kang K et al (eds) Proceedings of the 2000 Winter Simulation conference. Society for Modelling and Computer Simulation International, La Jolla, pp 840–848
7. Brade D (2003) A generalized process for the verification and validation of models and simulation results. Dissertation, Fakultät für Informatik, Universität der Bundeswehr München, Germany
8. American Institute of Aeronautics and Astronautics (2002) Guide: guide to the verification and validation of computational fluid dynamics simulations (AIAA G-077-1998(2002)), AIAA, Reston
9. Hemez FM (2004) The myth of science-based predictive modelling. In: Proceedings foundations '04 workshop for verification, validation and accreditation (VV&A) in the 21st century, Arizona State University, Tempe, Arizona, 13–15 October 2004. Report LA-UR-04-6829, Los Alamos National Laboratory, USA
10. Montgomery DC, Conard RG (1980) Comparison of simulation and flight-test data for missile systems. Simulation 34(2):63–72

11. Oberkampf WL (2007) Predictive capabilities in computational science and engineering. Presented at OASCR applied mathematics PI meeting, Lawrence Livermore National Laboratory, 22–24 May 2007. http://science.energy.gov/~/media/ascr/pdf/workshops-conferences/mathtalks/Oberkampf.pdf.Accessed 5 June 2015
12. Rasmussen CE, Ghahramani Z (2001) Occam's razor. In: Leen TK, Dietterich TG, Tresp V (eds) Advances in Neural Information Processing Systems 13, Papers from Neural Information Processing Systems (NIPS) 2000, Denver CO, MIT Press, Cambridge MA, pp 294–300
13. The Mitre Corporation (2014) Verification and validation of simulation models. In: Mitre systems engineering guide, pp 461–469. www.mitre.org/publications/technical-papers/the-mitre-systems-engineering-guide. Accessed 5 June 2015
14. Pasquier M, Duoba M, Rousseau A (2001) Validating simulation tools for vehicle system studies using advanced control and testing procedures. In: Proceedings 18th international electric vehicle symposium (EVS18), Berlin, Germany
15. SCS Technical Committee on Model Credibility (1979) Terminology for model credibility. Simulation 32(3):103–104
16. Murray-Smith DJ (1998) Methods for the external validation of continuous system simulation models: a review. Math Comput Model Dyn Syst 4(1):5–31
17. Eddy DM, Hollingworth W, Caro JJ et al (2012) Model transparency and validation. A report of the ISPOR_SMDM Modeling Good Research Practices Task Force-7. Med Decis Making 35(5):733–743
18. Oberkampf WL, Roy C (2010) Verification and validation in scientific computing. Cambridge University Press, Cambridge
19. American Society of Mechanical Engineers (ASME) Committee V&V40 verification and validation in computational modeling of medical devices and associated sub-groups. https://cstools.asme.org/csconnect/CommitteePages.cfm?Committee=100108782
20. American Society of Mechanical Engineers (ASME) Committee on verification and validation in computational modeling and simulation and associated sub-committees (V&V 10, V&V20, V&V30 and V&V40). https://cstools.asme.org/csconnect/CommitteePages.cfm?Committee=100003367. Accessed 5 June 2015
21. Modeling and Simulation Coordination Office (M&S CO). http://www.acq.osd.mil/se/org_msco.html. Accessed 5 June 2015
22. Kuhn DR, Chandramouli R, Butler RW (2002) Cost effective use of formal methods in verification and validation. Invited paper, Workshop on foundations for modeling and simulation (M&S) verification and validation (V&V) in the 21st century (Foundations '02 Workshop), US Department of Defense, Laurel, Maryland, 22–23 October 2002
23. Cook DA, Skinner JM (2005). How to perform credible verification, validation and accreditation for modelling and simulation, J Def Softw Eng 18(5):20–24
24. Eisenberg N, Federline M, Sagar B et al (1995) Model validation from a regulatory perspective, GEOVAL'94, validation through model testing, Proceedings NAE/SKI symposium, Paris, France, 11–14 October 1994. Nuclear Energy Agency, Organisation for Economic Co-operation and Development, Paris, pp 421–434

Chapter 3
Measures of Quality for Model Validation

3.1 Choice of Output Variables for Model Quality Assessment

In practice, simulation models usually have more than one output variable and generate large amounts of information. In considering measures of model quality, the choice of outputs must depend on the planned application and, traditionally, analysis of model quality has been based on many different measures. For example, in engineering applications these measures could involve simple features that are important for a specific application, such as the magnitude of the peak value found in the time history of a particular variable for a given set of initial conditions and inputs. In other cases it might be a frequency of oscillation or a rate of decay of an output variable during a transient. Alternatively, the model quality assessment could involve the complete time histories recorded for several chosen variables.

In many model validation studies reported in the literature, emphasis has been placed on the use of a combination of quantitative measures and graphical representations. Quantitative measures should not be applied in isolation since much additional insight can often come from graphical methods.

In general, validation measures have to be established at an early stage of the modelling process when validation requirements and the testing strategy are being drawn up. The measures must relate to the model requirements, both in terms of the variables of primary interest and the aspects of system performance that are judged to be most important. Measures of model quality should also take account of the types of data available from tests and experiments and the potential difficulties and costs of making measurements, both at the level of the complete system model and at the sub-model and component levels.

D.J. Murray-Smith, *Testing and Validation of Computer Simulation Models*,
Simulation Foundations, Methods and Applications,
DOI 10.1007/978-3-319-15099-4_3

3.2 Graphical Methods and Measures

Graphical methods are often based on plots of simulated variables (usually continuous and represented by a line) and observed or measured values (often discrete) against an independent variable (often time). Wherever possible, measured results should include some estimate of experimental errors. In some applications, where a number of different sensors are available and checks can be made of compatibility of measurements obtained in different ways, powerful techniques of data evaluation and reconstruction can be applied to obtain a best of measured variables. For example, Extended Kalman Filtering techniques have been found to be particularly useful for these data evaluation and reconstruction tasks and have been especially successful in the application of system identification techniques to fixed-wing aircraft and helicopters.

There are many textbooks available which can provide the background information necessary for estimating and interpreting experimental errors (see, e.g., [1]). In the case of data sets gathered primarily for purposes other than model validation, or data gathered in previous projects, the available documentation may not allow accurate estimation of measurement errors but, ideally, this should still be attempted. In terms of graphical methods expected errors may be shown as error bars superimposed on the nominal values found experimentally which are usually shown as discrete points.

Something that may be missed by less experienced observers is the fact that the deviation between simulated and measured values in a time-history type of record is the vertical separation between corresponding points on the graphs at the same instant of time and not simply the apparent distance between simulated and measured curves. A plot of the difference between the model and system responses versus time can avoid this difficulty and can usefully augment the separate time-history plots of the variables. A related issue can arise in cases where the problem is caused by simple time delays and it has been suggested that any quantitative comparison on a point by point basis should, in such situations, be based on an error measure that uses "visual" rather than vertical distances between the experimental and simulated plots [2]. This measure is claimed to be closer to "what the eye sees".

Another commonly used form of graph involves simulated values plotted against the corresponding measured or observed values. Ideally this plot should be a straight line at an angle of 45° to the axes. Deviations from the ideal are shown by the vertical distance between the points and the 45-degree line. Points above the 45-degree line are clearly over-estimated in the simulation while any points below the line are under-estimated.

Although the graphical presentation of information in this way is very easy to interpret in the case of a single variable, the task becomes more and more difficult as the number of variables being considered increases. Figure 3.1 shows a practical case involving two output variables. The application in this case is a helicopter simulation model [3]. It can be seen that, although there is relatively good

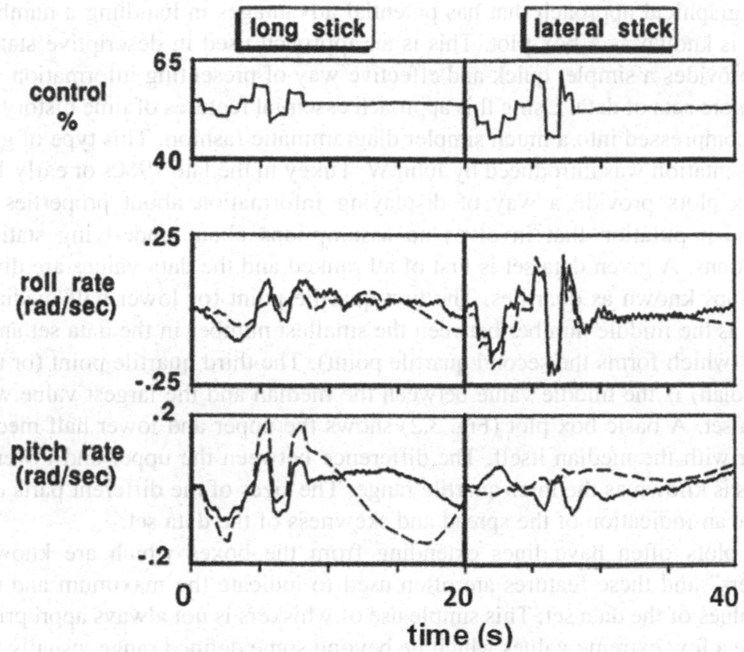

Fig. 3.1 Roll rate and pitch rate responses versus time from flight tests of a Bo-105 helicopter and from simulation using the DLR (German aerospace research laboratory) SIMH simulation model and DLR flight test data. Control inputs involve successive application of perturbations to the longitudinal and lateral stick positions as shown in the top records. The total period of each experimental record is approximately 24 s. Simulation results are shown by *dashed lines* and measured output responses in terms of roll rate and pitch rate are shown, in the two lower records, by the *continuous curves* in each case. Measured data have been subjected to pre-processing to check compatibility of the signals from different sensors, to correct subsequently for measurement errors and to evaluate the data quality. The original version of this figure was published by the Advisory Group for Aerospace Research and Development, North Atlantic Treaty Organisation (AGARD/NATO) in AGARD Advisory Report 280 'Rotorcraft System Identification', September 1991 [3]

agreement in terms of the roll rate variable, the difference between the measured and simulated pitch rate variables is much larger. The model in question represents the 6-degrees-of-freedom motion a helicopter.

Although subjective in their interpretation, graphical presentations such as these can be very useful in model validation and can complement quantitative measures for relatively simple cases involving a small number of plots. Problems of graphical comparison obviously become greater as the number of records is increased. For example, if the number of variables was to be increased so that there were now 20 or more measured outputs to be compared with the corresponding simulated quantities, the problem of interpreting the graphical information correctly would become much more demanding and other visualisation methods would be needed.

One graphical approach that has potential advantages in handling a number of records is known as a box plot. This is an approach used in descriptive statistics which provides a simple, quick and effective way of presenting information about one or more sets of data. Using this approach essential features of time history plots can be compressed into a much simpler diagrammatic fashion. This type of graphical presentation was introduced by John W. Tukey in the late 1960s or early 1970s [4]. Box plots provide a way of displaying information about properties of a statistical population that involves no assumptions about underlying statistical distributions. A given data set is first of all ranked and the data values are divided into groups known as quartiles. The first quartile point (or lower half median) is defined as the middle number between the smallest number in the data set and the median (which forms the second quartile point). The third quartile point (or upper half median) is the middle value between the median and the largest value within the data set. A basic box plot (Fig. 3.2) shows the upper and lower half medians, together with the median itself. The difference between the upper and lower half medians is known as the inter-quartile range. The sizes of the different parts of the box give an indication of the spread and skewness of the data set.

Box plots often have lines extending from the boxes which are known as "whiskers" and these features are often used to indicate the maximum and minimum values of the data set. This simple use of whiskers is not always appropriate if there are a few extreme values which lie beyond some defined range, usually taken as four times the inter-quartile range. Any data points outside that range are not included in the data used to form the diagram and are regarded as outliers. They are indicated on the box plot by individual dots or circles, as shown in Fig. 3.3.

In terms of model validation applications, box plots are most likely to be of value in displaying, in a concise fashion, the properties of data sets representing the

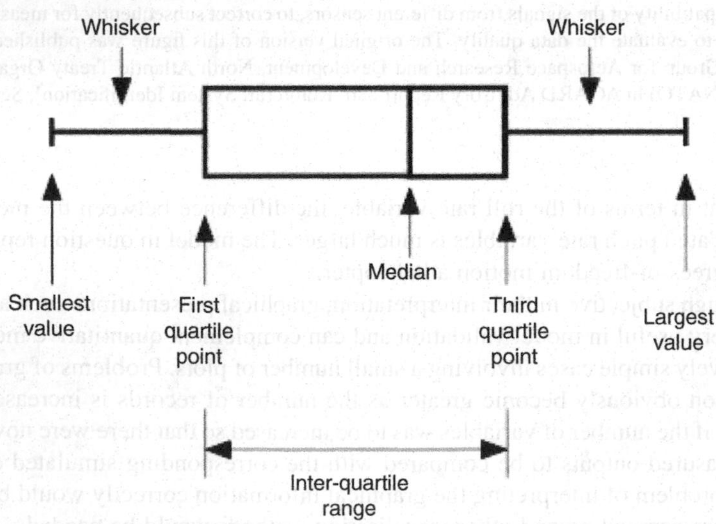

Fig. 3.2 Box plot with whiskers

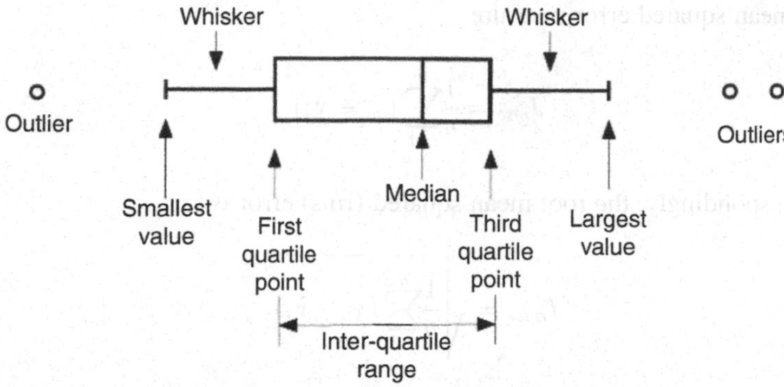

Fig. 3.3 Box plot with whiskers and outliers

differences between measured and simulated time histories for tests carried out under equivalent conditions. Instead of a simple error time-history plot formed by taking the difference between measured and model outputs at each sample time point, a box and whisker plot may provide a more effective way of presenting the information. This is especially true when several variables of the model and system are being used in the comparison.

Graphical output is often very important in "holistic" methods for the assessment of models. These are often based upon the opinion of someone who is very familiar with the behaviour of the real system being modelled. Expert opinion of this kind has to be based on the overall behaviour of a model compared with that of the real system and is likely to involve consideration of a number of different experimental cases and also several operating conditions.

3.3 Deterministic Quantitative Measures

3.3.1 Time-Domain Measures

Quantitative measures for system and model comparison are clearly important for simulation model validation. Our interest lies in measures that allow comparison of outputs from a deterministic computer simulation model with test measurements of corresponding variables from a real system for the same experimental conditions. Among the most used deviance measures are the mean absolute and mean-square errors. For the case of n sets of measured and simulated values for a single variable, the difference between observed values y_i and simulated values $\hat{y}_i$ is used to give the mean absolute error:

$$J_{mae} = \frac{1}{n}\sum_{i=1}^{n} |y_i - \hat{y}_i| \tag{3.1}$$

For the mean squared error measure:

$$J_{mse} = \frac{1}{n}\sum_{i=1}^{n} \left(y_i - \hat{y}_i\right)^2 \tag{3.2}$$

and, correspondingly, the root mean squared (rms) error is:

$$J_{rmse} = \sqrt{\frac{1}{n}\sum_{i=1}^{n} \left(y_i - \hat{y}_i\right)^2} \tag{3.3}$$

In many situations it is useful to include some form of weighting function so that errors arising in specific sections of the time history can be given special emphasis. For example, one such measure of fit is provided by the weighted rms error:

$$J_{wrms} = \sqrt{\frac{1}{n}\sum_{i=1}^{n} \left(y_i - \hat{y}_i\right)^T w_i \left(y_i - \hat{y}_i\right)} \tag{3.4}$$

where w is a weighting factor and the superscript ^{T} indicates the transpose.

As with the other measures mentioned above, this can be extended to the case where there are n_p separate output records (perhaps corresponding to the state variables of the model of the system under investigation or a sub-set of the state variables). If each record has n sampled values, the rms error is given by:

$$J_{rms} = \sqrt{\left(\frac{1}{nn_p}\right)\sum_{i=1}^{n} \left((\mathbf{y} - \hat{\mathbf{y}})\right)^T \mathbf{W} \left(\mathbf{y} - \hat{\mathbf{y}}\right)} \tag{3.5}$$

One obvious difficulty with the measures defined in (3.1), (3.2), (3.3), (3.4), and (3.5) is that they depend on the units of the variables being compared. Measures of this kind, determined for a number of different variables of a given model and system, cannot be compared directly because they are likely to involve different units. One approach that avoids this problem involves the use of relative error measures and an example of this is the mean absolute percent error, given by:

$$J_{map} = \frac{100}{n}\sum_{i=1}^{n} \frac{|y_i - \hat{y}_i|}{|y_i|} \tag{3.6}$$

Although this has the advantage of avoiding problems associated with units, relative error measures of this kind are inapplicable if any of the observed values happens to equal zero and difficulties may also arise for situations where measured values are small. Such measures may, however, be useful in dealing with model variables that can only take positive values, such as concentrations or measures of energy.

An obvious disadvantage, which applies to all of the above measures, is that they can show high sensitivity to single extreme values. Thus, as with the box diagram approach discussed in Sect. 3.2, outliers may have to be identified and their values may be excluded from the measure.

A convenient measure that has been used for model validation applications in a variety of application areas is Theil's Inequality Coefficient (TIC), which is defined, in the case of a single variable, as:

$$J_{TIC} = \frac{\sqrt{\frac{1}{n}\sum_{i=1}^{n}(y_i - \hat{y}_i)^2}}{\sqrt{\frac{1}{n}\sum_{i=1}^{n} y_i^2} + \sqrt{\frac{1}{n}\sum_{i=1}^{n} \hat{y}_i^2}} \tag{3.7}$$

This measure provides values that lie between zero and unity and this can be an advantage compared with some of the simpler measures considered above. A value of J_{TIC} of zero indicates a model capable of perfect predictions for the variables considered and the range of measured data used in the comparison. A value of J_{TIC} equal to unity indicates no predictive capability. In general terms, a J_{TIC} value of the order of 0.2–0.3 is regarded as indicating satisfactory predictive agreement.

Theil's Inequality Coefficient can be extended, without difficulty, to cases in involving n_p sets of variables. However, the obvious difficulty with a measure of this kind is that one then only has one number describing the overall quality of a model which may include a number of different variables that are of interest. If the measure shows that the model, overall, has poor predictive capabilities one cannot see from the use of this measure alone where the problems may lie.

Other scaled measures have also received attention for applications involving comparisons of model and measured system time histories. An example is a normed root-mean-square output error measure which is similar in most respects to the TIC but involves normalisation based only on the sum of the squared values of experimental response samples [5].

Problems can easily arise if too much emphasis is placed on measures such as those discussed in this section. The importance of this may be seen from the simple example presented in Fig. 3.4 which is adapted from a case presented by Jachner et al. [6]. This shows three curves where the curve labelled "Data" represents a short section of the response of a simulated second-order linear system with an additional small time delay. This is intended to represent samples from a measured response obtained during an experimental step response test. The second curve, shown as "Model 1", is the response of a simulation model which, in fact, is identical to the system that generated the top-most curve, but without the pure time delay. The third curve, shown as "Model 2", is the response of a much simpler model involving only a pure integrator with a gain factor and an initial condition. Visual inspection of the three curves would immediately suggest that the "Model 1" response matches the original more closely than the "Model 2" response. However, evaluation of the

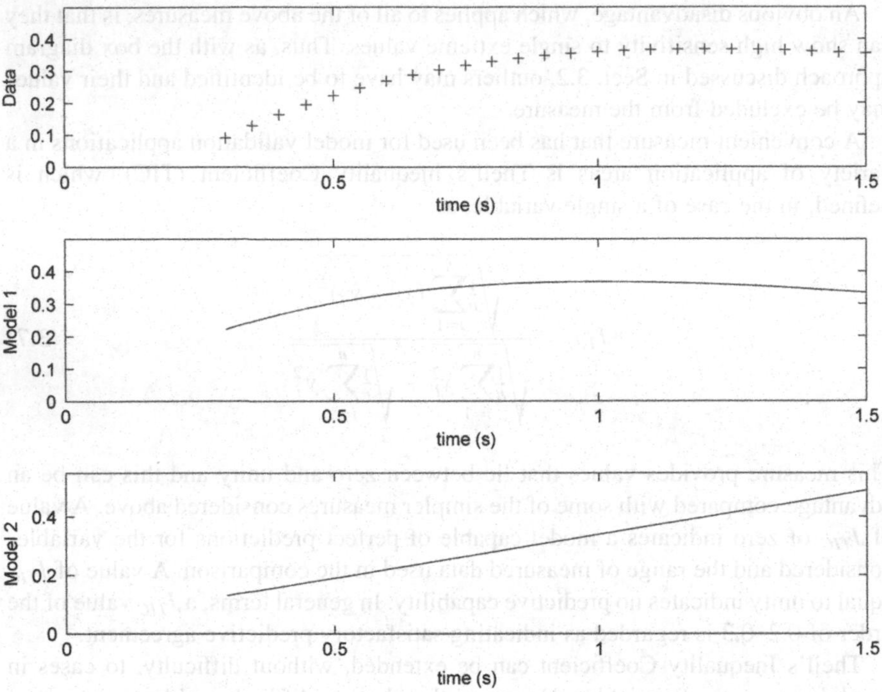

Fig. 3.4 Two model responses compared with "measured" data. The data samples were generated using a simple second-order under-damped linear model with an additional small time delay. The first model response (labeled "Model 1") was obtained from a model which was identical to that used to generate the data but with no time delay. The second model ("Model 2") involved only an integrator, with a gain factor and an initial condition

mean-squared error values and the Theil Inequality Coefficient values for the differences between the data and the model values does not support this finding. These measures, evaluated for the sample points in the data record, produce values which are significantly smaller for Model 2 than for Model 1. This would suggest that Model 2 is in some way superior to Model 1. However, we know that the lack of fit for Model 1 is due simply to an error in the model structure through omission of the time delay factor. If the time delay was taken into account fully, the fit between Model 1 and the data record would be perfect and no amount of adjustment of parameters could allow Model 2 to achieve a comparable situation. This example has, of course, been chosen as a rather extreme case but the message should be clear. Simple numerical measures should not be used in isolation when assessing models and other evidence should always be taken into account, especially information presented in graphical form.

3.3.2 Frequency-Domain Measures and Comparisons

In some specific fields, such as structural dynamics, frequency-domain methods are often used to compare responses from finite element models with measurements. This can be carried out graphically using a subjective approach to the assessment of differences between frequency response curves. Approaches of this kind have links with frequency response methods of system identification which are discussed in Chap. 5. Such approaches have obvious limits, but do provide clear physical evidence in terms of the parts of the frequency range where model deficiencies are greatest and thus can assist in terms of re-assessment of a model. This is discussed in Sect. 7.3, where an application of this type of approach (see also [7]) is outlined for the development of lumped parameter models of large hydraulic pipelines.

Although graphical techniques are important, there has also been interest in the development of quantitative measures for comparison of model frequency responses. One specific frequency response function (FRF) that has been proposed as a measure of model quality is the Frequency Response Assurance Criterion (FRAC) [8].

The FRAC involves a measure:

$$J_{FRAC} = \frac{\left|\left(H_{Xij}(\omega)\right)^{H}\left(H_{Mij}(\omega)\right)\right|^{2}}{\left[\left(H_{Xij}(\omega)\right)^{H}\left(H_{Xij}(\omega)\right)\right]\left[\left(H_{Mij}(\omega)\right)^{H}\left(H_{Mij}(\omega)\right)\right]} \tag{3.8}$$

where $(H_{Xij}(\omega))$ is the frequency response function for experimental data at response coordinate i and excitation coordinate j and $(H_{Mij}(\omega))$ is the corresponding function for the model. The superscript H used in (3.8) indicates the complex conjugate (Hermitian) transpose. For frequency response functions that are identical the value of J_{FRAC} is unity while a value of zero would indicate completely uncorrelated response functions.

The paper by Heylan and Lammens [8] provides an interesting illustration of the use of frequency-domain methods, including a practical application to a test rig. Measured and analytically derived frequency response functions are compared over a specified frequency range for a relatively coarse finite element model which is then updated in a series of iterations until an acceptable level of agreement between the frequency response functions is obtained. Results showed resonant frequency differences of less than 4 % and mode shape agreement of 92 % for the first four modes, as indicated by a Modal Assurance Criterion suggested by Allernang [9]. Although developed for structural applications, this approach is of potential interest for verification and validation of finite element models in other areas of application. Further information about relatively recent developments in frequency-domain methods applied to structural dynamics problems may be found in theses by Grafe [10] and by Kershen [11].

Another important type of frequency-domain measure that can be useful in system modelling from measured response data is the "ordinary coherence function" (or "magnitude squared" coherence) between two signals (or data sets) $u(t)$ and $y(t)$. Essentially this function gives an estimate of the power transfer between the two signals and thus can allow causality between the input and output signals in a given system to be assessed. It is a real-valued function defined as:

$$\gamma^2(\omega) = \frac{|G_{uy}(\omega)|^2}{G_{uu}(\omega)G_{yy}(\omega)} \tag{3.9}$$

where $G_{uu}(\omega)$ is the auto-spectrum of the input $u(t)$, $G_{yy}(\omega)$ is the auto-spectrum of the output $y(t)$ and $G_{uy}(\omega)$ is the cross-spectral density from input u to output y.

The ordinary coherence function is a measure of linear dependence between an input and an output response as a function of frequency ω. Values always lie between zero and unity and, for an ideal single-input single-output linear system, it can be shown, through simple analysis, that the coherence will always be one. However, in real physical systems, measurement noise and nonlinearity are always present and a single-input single-output linear description is seldom adequate. An ordinary coherence function value which is less than unity usually indicates that the relationship between the input $u(t)$ and the output $y(t)$ is not linear, or that significant noise is present or that $y(t)$ incorporates output components from inputs other than $u(t)$. Of course, if the ordinary coherence function is found to have a value of zero we know that the two signals are unrelated.

It should be noted that if the system being considered is non-stationary, analysis based on the ordinary coherence function may be inappropriate and the concept of coherence has been extended to deal with such situations using time-frequency distributions (see e.g. [12]).

3.4 Statistical Measures

In addition to the deterministic type of measures discussed in Sect. 3.3, statistically-based methods and measures may also be applied in the development and application of dynamic models. Such methods are most commonly found in applications that involve systems that have significant random disturbances or display other stochastic behaviour. For example, statistically-based methods have been proposed for comparing time series. One approach involves fitting an appropriate stochastic model to each of the time series and then comparing the two models. If the models are the same it is clear that the two time series are the same and the greater the difference between the two models the greater the difference between the two time series. The type of description used often takes the form of an autoregressive integrated moving average model. Difficulties with this approach arise when the available time series representing real measured response data is short or where unexpected pure time delays are present in the measurements. In some process type applications, simple quantities such as the deviation from the mean (the "divergence") [13] can be useful, along with the deviation in spectral signal energy in the model and system measurements. The box-plot approach outlined in Sect. 3.2 represents another very simple, but effective, way of presenting statistical information in the context of data analysis and time series. That approach based on descriptive statistics. is one of many statistical techniques used in the investigation of issues of model quality and model validation. Further information may be found within many textbooks dealing with time series analysis methods (see, e.g. [14, 15]).

In the case of applications involving control system design, much use has been made of system identification methods involving deterministic models with an additional "noise" model to account for the effects of measurement noise and uncertainties. A related approach, which has also been used in control system design applications, is outlined by Correa and Postlethwaite [16]. In that work the issue of how to obtain information about model uncertainty from system identification results is approached using the idea of a "nominal" model plus a likely model perturbation, rather than performing the control system design using the nominal model on its own.

3.5 Visualisation Techniques

3.5.1 Polar Diagrams

One approach to visualisation, which is recognised as being useful in cases involving relatively complex models, is based on a number of key measured system or sub-system quantities which are represented as points on radial lines on an appropriately scaled polar diagram, as shown in Fig. 3.5, where continuous lines represent measured quantities and the dashed lines are equivalent quantities from a model. A polygon of model results and a polygon of corresponding measurements on the same polar diagram then provides an indication of the fidelity of the model. Polygonal shapes that are very similar may indicate that the model is broadly suitable for the application, while any areas requiring further investigation are also highlighted. This graphical approach to the presentation of information about a system and model is considered further in Chap. 10 in the case study concerned with helicopter flight mechanics modelling.

Diagrams of this kind also provide a convenient basis for sensitivity analysis since distortion of the model polygon following a specific change provides a clear indication of sensitivities and interactions between parameters [17, 18]. Visualisation methods of this kind lend themselves to image processing techniques for quantification of differences between the model and system polygons and several approaches have been considered [19, 20]. Such graphical comparisons make it clear which aspects of a system are represented accurately and which areas require further investigation. Although developed independently for the purposes of model

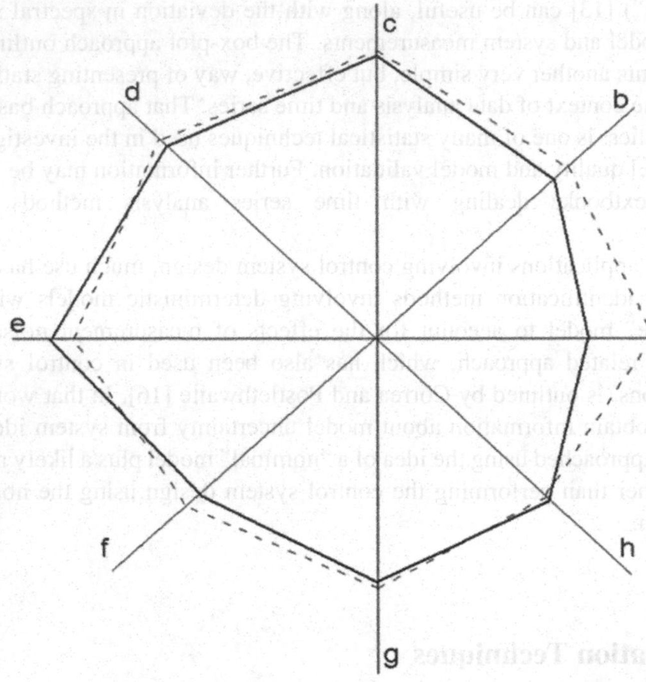

Fig. 3.5 Polar type of display with eight different quantitative measures from the system (shown by points connected by *dashed lines*) and the corresponding measures from the simulation model (shown by points connected by *solid lines*) (From [17])

validation, diagrams of this kind are recognised as being similar to Kiviat diagrams which are commonly used for visualisation of performance metrics for computer software and hardware.

These polar diagrams are clearly applicable to problems in many areas where there is a need to display results from several outputs (see, e.g. [21]). This approach to visualisation is also very flexible in terms of the system and model comparisons that can be made and can be used with any deterministic measures of performance, such as the size of a system response overshoot or the frequencies of observed oscillations. One further example of the use of polar diagrams can be found in the work of Kammel et al. [13] where a polar diagram is used to demonstrate the effect of model parameters on time-domain responses and evaluate possible benefits of process optimisation.

Recently, there has been a trend away from accuracy-centred assessment of models towards forms of assessment that are based more broadly on overall model fidelity using a combination of objective measures and more subjective assessments, including the use of animation methods. Other relevant discussions may be found in many sections of the book edited by Cloud and Rainey [22], in the textbook on the theory of modelling and simulation by Zeigler et al. [23] and in the work of Brade and Köster [24] and Brade et al. [25].

References

1. Hughes IG, Hase TPA (2010) Measurements and their uncertainties: a practical guide to modern error analysis. Oxford University Press, Oxford
2. Marron JS, Tsybakerov AB (1995) Visual error criteria for qualitative smoothing. J Am Stat Assoc 90:499–507
3. Advisory Group for Aerospace Research and Development, North Atlantic Treaty Organisation (AGARD/NATO) in AGARD Advisory Report 280 'Rotorcraft System Identification', September 1991
4. Tukey JW (1977) Exploratory data analysis. Addison Wesley, Reading
5. Knudsen M (2006) Experimental modelling of dynamic systems: an educational approach. IEEE Trans Educ 49(1):29–38
6. Jachner S, van den Boogaart KG, Petzoldt T (2007) Statistical methods for the qualitative assessment of dynamic models with time delay (R Package qualV). J Stat Softw 22(8)
7. Bryce GW, Foord TR, Murray-Smith DJ et al (1976) Hybrid simulation of water-turbine governors. In: Crosbie, RE, Hay JL (eds) Towards Real-Time Simulation (Languages, models and systems) Part 1, Simulation Councils Proceedings Series, 6(1), Chapter 6, pp 35–44. Simulation Councils, La Jolla, CA
8. Heylen W, Lammens S (1996) FRAC: a consistent way of comparing frequency response functions. In: Proceedings of conference on identification in engineering systems, Swansea, pp 48–57
9. Allernang RJ (1980) Investigating some multiple input/output frequency response function experimental model analysis techniques. PhD dissertation, University of Cincinnati, USA
10. Grafe H (1998) Model updating of large structural dynamics models using measured response functions. PhD dissertation, Imperial College, London
11. Kershen G (2002) On the model validation in non-linear structural dynamics. Doctoral dissertation, University of Liege, Belgium
12. White LB, Boashash B (1990) Cross-spectral analysis of non-stationary processes. IEEE Trans Inf Theory 36(4):830–835
13. Kammel G, Voigt HM, Neβ K (2005) Development of a tool to improve the forecast accuracy of dynamic simulation models for the paper process. In: Kappen J, Manninen J, Ritala R (eds) Proceedings of model validation workshop. VTT Technical Research Centre, Finland, Espoo
14. Chatfield C (2003) The analysis of time series: an introduction, 6th edn. Chapman and Hall/CRC, Boca Raton
15. Brockwell PJ, Davis RA (2002) Introduction to time series and forecasting, 2nd edn. Springer, New York
16. Correa GO, Postlethwaite I (1986) Model uncertainty and linear system identification. In: Skelton RE, Owens DH (eds) Proceedings of the IFAC workshop on model error concepts and compensation, Boston, 1985, pp 115–120, Pergamon, Oxford
17. Smith MI, Murray-Smith DJ, Hickman D (2007) Mathematical and computer modeling of electro-optic systems using a generic modeling approach. J Def Model Simul 4(1):3–16
18. Smith MI, Murray-Smith DJ, Hickman D (2007) Verification and validation issues in a generic model of electro-optic sensor systems. J Def Model Simul 4(1):17–17
19. Grant A, Murray-Smith DJ (204) Polygon imaging methods in plant monitoring and model validation applications. In: Proceedings Advanced Engineering Design 2004 Conference, Glasgow, September 2004, Paper C1.07. Orgit, Prague, Czech Republic
20. Grant AGN (2006) Data to polygon transformation and comparison: a technique for model validation and related applications. PhD dissertation, University of Glasgow, UK
21. Murray-Smith DJ (2012) Modelling and simulation of integrated systems in engineering: issues of methodology, quality, testing and application. Woodhead Publishing, Cambridge
22. Cloud DJ, Rainey LB (eds) (1998) Applied modeling and simulation: an integrated approach to development and operation. McGraw-Hill, New York

23. Zeigler BP, Praehofer H, Lim TG (2000) Theory of modeling and simulation, 2nd edn. Academic, San Diego
24. Brade D, Köster, A (2001) Risk-based validation and verification levels definition. In: Proceedings of the European simulation interoperability workshop, London, UK2001, Simulation Interoperability Standardization Organization
25. Brade D, Maguire R, Lotz H-B (2002) Arguments-based credibility levels. In: Proceedings of the SISO spring simulation interoperability workshop 2002, Orlando

Chapter 4
Sensitivity Analysis for Model Evaluation

Sensitivity analysis is a well-established approach for investigating how much a given variable within a model depends on quantities within that model (see, e.g. [1]). Sensitivity analysis of a complete system model can help to identify potential problem areas which may give rise to high dependencies of outputs on specific sub-system elements. This is of fundamental importance for system identification procedures intended to provide reliable parameter estimates at the model calibration stage and also for validation and upgrading of models.

The process may involve investigation of the effect of changes of model structure or changes of parameters of the model, or both of these. Within most application areas the emphasis is on the analysis of the sensitivity to parametric changes, but structural sensitivity information can provide additional insight in some situations, such as the investigation of modelling assumptions. In cases where parameter changes interact with issues of model structure to produce changes in overall behaviour of a model the sensitivity analysis is termed "singular perturbation" analysis (see, e.g., [2]).

4.1 An Introduction to Sensitivity Analysis Methods

Investigating which parameters influence model behaviour most is potentially important for the iterative development and refinement of models. Gradient methods of optimisation make direct use of parameter sensitivity information and, in engineering applications, parameter sensitivity information is essential in order to understand the effects of component tolerances. It also allows predictions to be made about how the performance of an engineering system may change with ageing or alterations of the environment. For the purposes of sensitivity analysis the form of model may be continuous or discrete in nature, but this is of little

D.J. Murray-Smith, *Testing and Validation of Computer Simulation Models*,
Simulation Foundations, Methods and Applications,
DOI 10.1007/978-3-319-15099-4_4

significance in terms of the objectives of sensitivity analysis and the insight that may come from its application.

One interesting issue relates to the stage of model development at which sensitivity analysis can be usefully applied. Clearly it is not very helpful if the benefits of being able to assess parameter sensitivities are only achievable once the model has been subjected to validation tests and has been found to be fit for purpose. On the other hand, it is also clear that sensitivity information gathered before the model has been fully validated are of doubtful value. Fortunately, it has been found, through experience, that parameter sensitivity analysis can often be used to establish the relative importance of different parameters of a model even when that model is far from optimal in terms of its parameter values. This means that it may often be possible to produce meaningful results from sensitivity analysis during the formulation and development stages of a modelling project, provided the structure of the model is appropriate.

An illustration of this has been provided by Miller [3] in an account of an ecological modelling study involving a model with 6 compartments and 18 associated parameters. The nominal values of each parameter of the model were multiplied by a factor $(1 + \delta)$ where δ in each case was randomly chosen from a population with normal distribution and zero mean and with a given standard deviation. Two cases were considered involving standard deviations of 0.5 and 1.0, corresponding to 50 % or 100 % differences from the nominal parameter values. If the application of this procedure led to a change of sign of a specific parameter the selected δ value was discarded and a new value was drawn. Results obtained by Miller showed that even with these large errors in parameter values the four most important parameters of the model could still be distinguished when sensitivity analysis was carried out with parameters perturbed in this way. Similarly, the five least important parameters for the nominal model could be distinguished with a perturbation of parameters of 50 %. With 100 % change only the three least important parameters were preserved. A more drastic test in which all parameters of the model were set to a fixed value of 0.01 showed that the four most sensitive parameters for the nominal model remained in the group of four most sensitive parameters. However, this was not true of the least sensitive parameters, where only two of the five least significant quantities within the nominal model were included in the corresponding set for the perturbed model. All of this suggests that it may be possible to make use of sensitivity analysis methods early in the development of a model and establish which parameters are likely to be a critical importance and which may be considered as fixed quantities or discarded completely.

4.2 Sensitivity Functions

The concept of a sensitivity function (sometimes termed a sensitivity "coefficient") is central to all methods of sensitivity analysis. For models involving continuous variables in the time domain, sensitivity functions may be defined using a Taylor series expansion in terms of the quantity that is varying (often a parameter of the given model). If $y(t, q_0)$ is the output of the model for a given set of inputs and this output variable is a function of time, t, and also of a parameter, q, the response for a slightly altered parameter value $q = q_0 + \Delta q$ is given by:

$$y(t, q_0 + \Delta q) = y(t, q_0) + \frac{\partial y}{\partial q}\Delta q + \frac{1}{2!}\frac{\partial^2 y}{\partial q^2}(\Delta q)^2 + \cdots \tag{4.1}$$

and if Δq is small

$$y(t, q_0 + \Delta q) \approx y(t, q_0) + \frac{\partial y}{\partial q}\Delta q \tag{4.2}$$

The partial derivative $\frac{\partial y}{\partial q}$ is known as the first-order output sensitivity function.

The linearisation resulting from truncation of the series after the second term means that the superposition principle can be applied and this allows simultaneous effects of changes of a number of parameters to be considered. Sensitivity functions $\frac{\partial x}{\partial q}$ for state variables x of a state-space type of model are known as "trajectory sensitivity functions" and are written as $\frac{\partial \mathbf{x}}{\partial q}$ where $\boldsymbol{x}$ is the vector of state variables. As in the case of output sensitivity, the trajectory sensitivity is evaluated for perturbations in the parameter value q about a specific value q_0.

The "relative" sensitivity function provides a useful measure which allows the influence of a number of different parameters to be compared. Where the comparisons involve one specific model variable x and number of different parameters $\boldsymbol{q}$, an appropriate measure of relative sensitivity is $q_i \frac{\partial x}{\partial q_i}$. For cases with more than one variable of interest a dimensionless form of relative sensitivity measure $\frac{q_i}{x_{mj}}\frac{\partial x_j}{\partial q_i}$ is more appropriate. Here the quantity x_{mj} represents some appropriate measure of the variable $x_j(t, q_0)$, such as a peak value or mean square value. One example of the use of relative sensitivity functions are the sensitivity measures used by Miller [3] in his compartmental model mentioned in Sect. 4.1.

4.2.1 Parameter Perturbation Methods of Sensitivity Analysis

Rearranging (4.2) allows the sensitivity function to be written as:

$$\frac{\partial y}{\partial q_0} \approx \frac{y(t, q_0 + \Delta q) - y(t, q_0)}{\Delta q} \tag{4.3}$$

Sensitivity functions can thus be found from a finite difference approximation involving repeated simulation runs and selection of an appropriate parameter perturbation, Δq. The accuracy depends on the magnitude of the perturbation and tests may have to be repeated using several trial values of Δq to find a suitable perturbation level. Separate calculations are required for each parameter being considered.

Difficulties in applying this approach include the fact that it requires separate simulation runs for the situation before and after the perturbation, which may have computational costs if the models are complex. There are also obvious issues of accuracy. The process of taking the difference between the two solutions, each of which has inherent numerical errors ε_1 and ε_2, leads to an equation:

$$\frac{\partial y}{\partial q_0} \approx \frac{y(t, q_0 + \Delta q) - y(t, q_0)}{\Delta q} + \frac{\varepsilon_1 - \varepsilon_2}{\Delta q} \tag{4.4}$$

Since the perturbation Δq must be small the error resulting from the term $(\varepsilon_1 - \varepsilon_2)$ $/\Delta q$ may be significant [1].

The relative sensitivity measures corresponding to (4.3) are:

$$q\frac{\partial y}{\partial q_0} \approx q\frac{y(t, q_0 + \Delta q) - y(t, q_0)}{\Delta q} \tag{4.5}$$

and

$$\frac{q}{y_m}\frac{\partial y}{\partial q_0} \approx \frac{q}{y_m}\frac{y(t, q_0 + \Delta q) - y(t, q_0)}{\Delta q} \tag{4.6}$$

where y_m is an appropriate reference measure of the variable $y(t, q_0)$.

4.2.2 Sensitivity Analysis Through the Use of Sensitivity Models

If a general nonlinear dynamic model is described by a set of equations:

$$f_i(\dot{x}_i; x_1, x_2, \ldots, x_n;\ u_1, u_2, \ldots u_r; t; q_0) = 0 \tag{4.7}$$

$$g_i(y_i; x_1, x_2, \ldots, x_n;\ u_1, u_2, \ldots u_r; t; q_0) = 0 \tag{4.8}$$

where $x_1, x_2, \ldots, x_n$ are state variables, $u_1, u_2, \ldots u_r$ are inputs and $y_1, y_2, \ldots y_m$ are output variables, it is clear that the sensitivity of this model to variation of the parameter q may be found by partial differentiation with respect to q to give a set of "sensitivity" equations. These may also be termed the "sensitivity model" which is sometimes also known as the "sensitivity co-system". These equations are of the form:

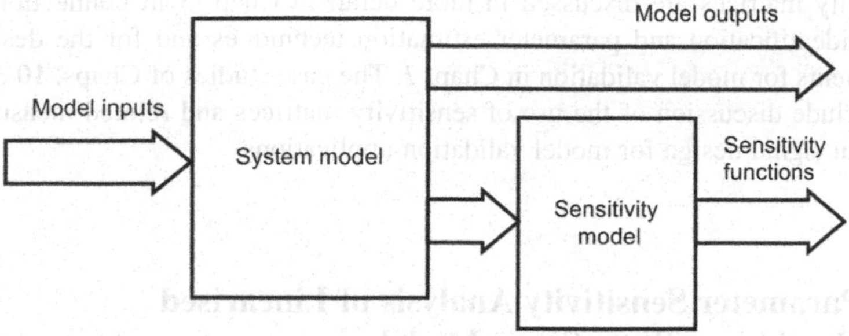

Fig. 4.1 Block diagram showing sensitivity model and process for generation of sensitivity functions for the general case of a multi-input multi-output nonlinear model

$$\frac{\partial f_i}{\partial \dot{x}_i}\frac{\partial \dot{x}_i}{\partial q_0}+\frac{\partial f_i}{\partial x_1}\frac{\partial x_1}{\partial q_0}+\ldots\ldots\ldots\ldots+\frac{\partial f_i}{\partial x_n}\frac{\partial x_n}{\partial q_0}+\frac{\partial f_i}{\partial q_0}=0 \quad (4.9)$$

$$\frac{\partial g_i}{\partial y_i}\frac{\partial y_i}{\partial q_0}+\frac{\partial g_i}{\partial x_1}\frac{\partial x_1}{\partial q_0}+\ldots\ldots\ldots\ldots+\frac{\partial g_i}{\partial x_n}\frac{\partial x_n}{\partial q_0}+\frac{\partial g_i}{\partial q_0}=0 \quad (4.10)$$

The sensitivity equations are, in general, linear ordinary differential equations with time-varying coefficients and have to be solved along with the equations of the model itself. This provides time histories of both the state variables and the relevant sensitivity functions. Figure 4.1 is a schematic diagram representing this procedure. In general, for each output of a given model the number of sensitivity models must equal the number of parameters of interest. However, in the case of time –invariant linear models the sensitivity models have structures that are very similar to the corresponding system model and methods are available for special cases that allow many sensitivity functions to be found simultaneously using a single sensitivity model.

Sensitivity models have initial conditions that are usually zero but, in some cases, such as those involving models with a variable structure, non-zero initial conditions can arise within the sensitivity model. This issue is considered in more detail by Frank [4].

The "sensitivity matrix" is a matrix of partial derivatives of model variables, such as state variables, with respect to parameters within a model. This matrix of sensitivity functions has particular importance for system identification and for model validation. For the case of n variables $y_1, y_2, \ldots y_n$ and p parameters $q_1, q_2, \ldots\ldots q_p$ the sensitivity matrix has the form:

$$\mathrm{X}=\begin{bmatrix}\frac{\partial y_1}{\partial q_1} & \cdots & \frac{\partial y_n}{\partial q_p}\\ \vdots & \ddots & \vdots\\ \frac{\partial y_n}{\partial q_1} & \cdots & \frac{\partial y_n}{\partial q_p}\end{bmatrix} \quad (4.11)$$

Sensitivity matrices are discussed in more detail in Chap. 5 in connection with system identification and parameter estimation techniques and for the design of experiments for model validation in Chap. 7. The case studies of Chaps. 10 and 11 both include discussion of the use of sensitivity matrices and related measures in test input signal design for model validation applications.

4.3 Parameter Sensitivity Analysis of Linearised Continuous State-Space Models

Consider a model of order n having m inputs and p outputs and described by a set of linearised equations (involving state variables $\boldsymbol{x}$, outputs $\boldsymbol{y}$ and inputs $\boldsymbol{u}$) which may be written in the standard state-space form as:

$$\dot{\boldsymbol{x}} = \boldsymbol{A}\boldsymbol{x} + \boldsymbol{B}\boldsymbol{u} \tag{4.12}$$

$$\boldsymbol{y} = \boldsymbol{C}\boldsymbol{x} + \boldsymbol{D}\boldsymbol{u} \tag{4.13}$$

Taking Laplace transforms and assuming zero initial conditions for the state variables gives:

$$s\boldsymbol{X}(s) = \boldsymbol{A}\boldsymbol{X}(s) + \boldsymbol{B}\boldsymbol{U}(s) \tag{4.14}$$

$$\boldsymbol{Y}(s) = \boldsymbol{C}\boldsymbol{X}(s) + \boldsymbol{D}\boldsymbol{U}(s) \tag{4.15}$$

Differentiating with respect to a parameter q, which can affect any or all of the coefficients of the matrices $\boldsymbol{A}$, $\boldsymbol{B}$, $\mathbf{C}$ and $\boldsymbol{D}$, the following set of equations are obtained:

$$s\frac{\partial \boldsymbol{X}}{\partial \boldsymbol{q}} = \boldsymbol{A}\frac{\partial \boldsymbol{X}}{\partial \boldsymbol{q}} + \frac{\partial \boldsymbol{A}}{\partial \boldsymbol{q}}\boldsymbol{X} + \boldsymbol{B}\frac{\partial \boldsymbol{U}}{\partial \boldsymbol{q}} + \frac{\partial \boldsymbol{B}}{\partial \boldsymbol{q}}\boldsymbol{U} \tag{4.16}$$

$$\frac{\partial \boldsymbol{Y}}{\partial \boldsymbol{q}} = \boldsymbol{C}\frac{\partial \boldsymbol{X}}{\partial \boldsymbol{q}} + \frac{\partial \boldsymbol{C}}{\partial \boldsymbol{q}}\boldsymbol{X} + \boldsymbol{D}\frac{\partial \boldsymbol{U}}{\partial \boldsymbol{q}} + \frac{\partial \boldsymbol{D}}{\partial \boldsymbol{q}}\boldsymbol{U} \tag{4.17}$$

In these equations the parameter q is a parameter of the system model and not of any of the inputs $\boldsymbol{U}(s)$ so that $\frac{\partial \boldsymbol{U}}{\partial \boldsymbol{q}}$ is zero. A solution of the sensitivity equations may thus be found using a simulation block diagram such as that in Fig. 4.2.

There are structural similarities between the system model and the sensitivity model and these can be shown to provide important practical computational benefits in the case of linear time-invariant models. However, in nonlinear or time-varying models, the link between the model structure and the form of the sensitivity equations is more complicated, although some similarities can still be found. The complexity then depends on the parameter of interest and whether or not

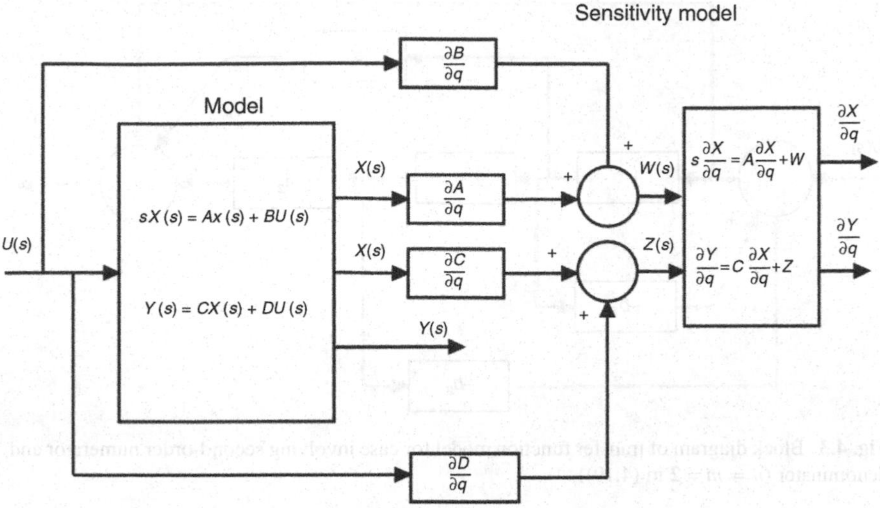

Fig. 4.2 Block diagram for sensitivity model for case of system model in linear state-space form

it is a parameter associated with a nonlinearity. Issues relating to sensitivity analysis of with nonlinear models are discussed in greater detail by Frank [4].

4.4 Parameter Sensitivity Analysis of Continuous Transfer Function Models

Sensitivity analysis of transfer function models described by linear ordinary differential equations with constant coefficients can be handled using the "sensitivity points" method, developed by Kokotović [5] and further developed by Wilkie and Perkins for models in state-space form [6].

For small changes of a parameter q, the sensitivity of a single-input single-output linear system with input $u(t)$ and output $y(t)$, described by a transfer function $G(s)$, may be found by partial differentiation:

$$\frac{\partial Y(s)}{\partial q_i} = \frac{\partial G(s)}{\partial q_i} U(s) = \frac{1}{G(s)} \frac{\partial G(s)}{\partial q_i} Y(s) \tag{4.18}$$

This shows that, in terms of Laplace transformed variables, the sensitivity function $\frac{\partial y(t))}{\partial q_i}$ can be found by applying the system output $y(t)$ as input to a sensitivity model with transfer function $\frac{1}{G(s)} \frac{\partial G(s)}{\partial q_i} Y(s)$. Kokotović has shown [5] that, for linear models, a sensitivity model may be represented by a multi-loop

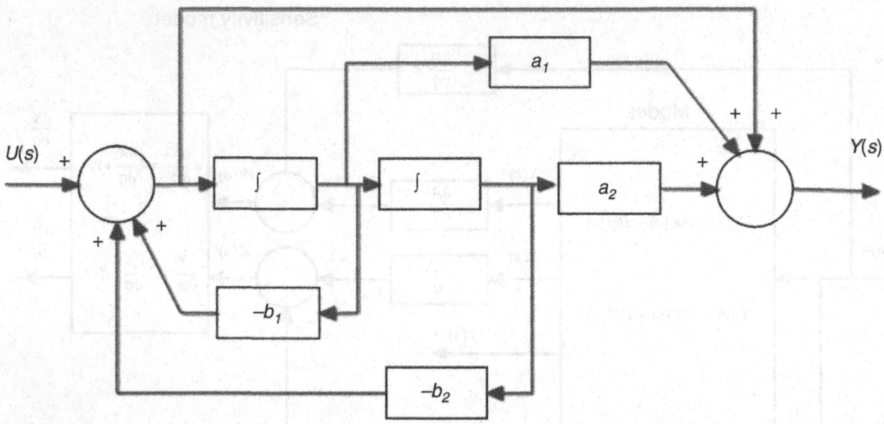

Fig. 4.3 Block diagram of transfer function model for case involving second-order numerator and denominator ($n = m = 2$ in (4.19))

feedback structure. As an example consider a transfer function of order n of the form:

$$G(s) = \frac{K\sum_{0}^{m} a_i s^{m-i}}{\sum_{0}^{n} b_j s^{n-j}} = \frac{K\sum_{0}^{m} a_i s^{m-n-i}}{\sum_{0}^{n} b_j s^{-j}} \tag{4.19}$$

where $a_0 = b_0 = 1$.

For the output $Y(s)$ it may then be shown that:

$$\frac{\partial Y(s)}{\partial b_r} = \frac{-s^{-r}Y(s)}{\sum_{0}^{n} b_j s^{-j}} \tag{4.20}$$

Figure 4.3 is a block diagram model for the case of $m = n = 2$ and the corresponding sensitivity model block diagram for the denominator coefficients is shown in Fig. 4.4. The points in the sensitivity model giving the sensitivity functions are the outputs of each integrator block. The sensitivity functions for all the coefficients of the denominator of $G(s)$ may thus be found simultaneously.

For the numerator coefficients

$$\frac{\partial Y(s)}{\partial a_r} = \frac{Ks^{m-n-r}U(s)}{\sum_{0}^{n} b_j s^{-j}} \tag{4.21}$$

Figure 4.5 shows the sensitivity model for the numerator coefficients for the model of Fig. 4.3. Here the integrator outputs for the system model $G(s)$ (i.e. the state variables) are the sensitivity functions for the numerator coefficients.

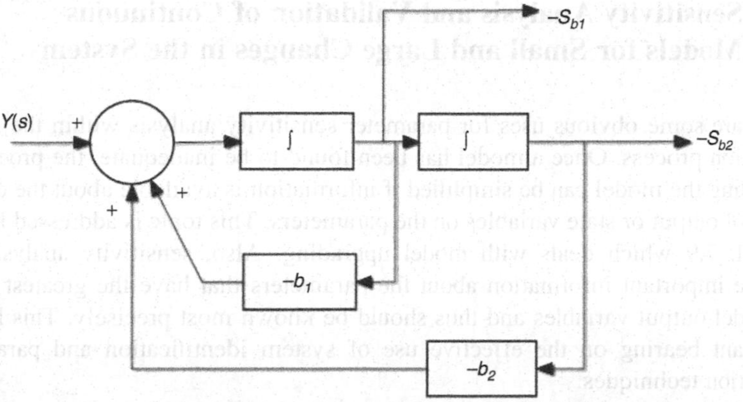

Fig. 4.4 Sensitivity model for generation of sensitivity functions for denominator coefficients for the model of Fig. 4.3

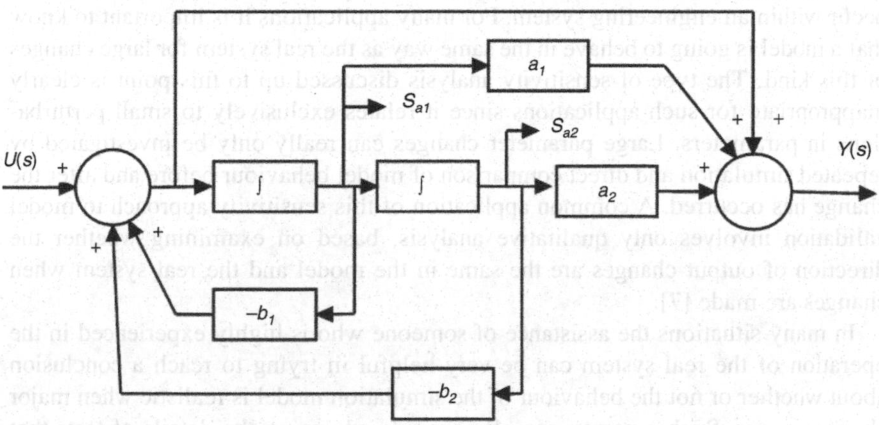

Fig. 4.5 Generation of sensitivity functions for numerator coefficients for second-order model of Fig. 4.3

A simulation that generates, simultaneously, the sensitivity functions of the system output with respect to all the numerator and denominator coefficients is easily established by combining together the results of (4.20) and (4.21). Since the output of each integrator block in Fig. 4.3 is a state variable, the sensitivity of each state variable to each parameter is directly available from a single sensitivity model for a system model in the given form [6]. It may also be shown that for multi-input linear systems with p inputs $2p-1$ nth order sensitivity models are needed (at most) to find all of the sensitivity functions of thestate variables [6].

4.5 Sensitivity Analysis and Validation of Continuous Models for Small and Large Changes in the System

There are some obvious uses for parameter sensitivity analysis within the model validation process. Once a model has been found to be inadequate, the process of upgrading the model can be simplified if information is available about the dependence of output or state variables on the parameters. This topic is addressed further in Sect. 7.9 which deals with model upgrading. Also, sensitivity analysis can provide important information about the parameters that have the greatest effect on model output variables and thus should be known most precisely. This has an important bearing on the effective use of system identification and parameter estimation techniques.

Another aspect of sensitivity analysis relates to the response of the model to large changes and this can also be important when tests of overall model credibility are being considered. This may involve sudden changes of the structure of the system or large changes of a parameter and could, for example, arise when faults occur within an engineering system. For many applications it is important to know that a model is going to behave in the same way as the real system for large changes of this kind. The type of sensitivity analysis discussed up to this point is clearly inappropriate for such applications since it relates exclusively to small perturbations in parameters. Large parameter changes can really only be investigated by repeated simulation and direct comparison of model behaviour before and after the change has occurred. A common application of this sensitivity approach to model validation involves only qualitative analysis, based on examining whether the direction of output changes are the same in the model and the real system when changes are made [7].

In many situations the assistance of someone who is highly experienced in the operation of the real system can be very helpful in trying to reach a conclusion about whether or not the behaviour of the simulation model is realistic when major changes occur. Such a person can also provide advice on the details of tests that should be performed. Again, this aspect of sensitivity analysis is considered further in the Chap. 7 where it forms part of the discussion of face validation methods.

4.6 Sensitivity Analysis of Discrete-Event and Hybrid Models

When dealing with discrete-event simulation models, sensitivity calculations are most often approached using repeated simulation runs and differencing techniques. Linear models described by difference equations, such as those that arise in modelling digital processors, can be handled using sensitivity model concepts similar to those outlined above.

As outlined in Chap. 1, hybrid systems involve both discrete variables and continuous variables. In hybrid systems these interactions are so significant that the continuous and discrete elements cannot be decoupled conveniently and must be analysed together. One of the commonest situations of this kind arises in hybrid systems with an embedded digital processor and associated analogue-to-digital (ADC) and digital-to-analogue (DAC) converters. In such cases it is common to discretise the continuous system sub-models using a sampling interval which is the same as the sampling interval of the ADC and DAC converters and use sampled-data representations within an overall discrete-time model. The processes of sensitivity analysis are then basically similar to the processes applied in continuous system simulation models.

Some other types of application that are regarded as hybrid systems for the purposes of system simulation may involve highly nonlinear elements such as relays or other kinds of switch. Hard nonlinearities of this kind present particular difficulties in terms of the use of general-purpose simulation tools in that the timing of events in the output record depend critically on these nonlinear elements of the model. A small error in the timing of events associated with hard nonlinearities, such as a switch, may produce significant errors in the final model output. The transitions occurring at nonlinear elements of this kind are known as "state events" and there are established methods for dealing with the problems of state event location (see e.g. [8–10]).

4.7 Discussion

The topic of sensitivity analysis is discussed further in the context of model validation in Chap. 7 and in some of the subsequent case studies chapters. The emphasis throughout is on the insight about aspects of model behaviour that can be obtained by using sensitivity analysis methods. The approach discussed in this chapter contrasts with some of the approaches to sensitivity analysis being proposed in recent years, which put more emphasis on automated methods (see, e.g. [11]). It is the belief of this author that, at least for the purposes of model validation, it is important to give the model user a central role within the sensitivity analysis process. Valuable insight may otherwise be lost. Automated procedures may have role for other types of application but are believed to be of limited value for the purposes of model validation.

References

1. Tomović R (1963) Sensitivity analysis of dynamic systems. McGraw-Hill, New York
2. Kokotović PV, O'Malley RE, Sannuti P (1976) Singular perturbation and order reduction in control theory – an overview. Automatica 12(2):123–132

3. Miller DR (1974) An experiment in sensitivity analysis on an uncertain model. Simulation 23:101–104
4. Frank PM (1978) An introduction to system sensitivity theory. Academic, London
5. Kokotović PV (1964) Method of sensitivity points in the investigation and optimisation of linear control systems. Autom Remote Control 25:1670–1676
6. Wilkie DF, Perkins WR (1969) Generation of sensitivity functions for linear systems using low order models. IEEE Trans Autom Control 14:123–129
7. The Mitre Corporation (2014) Verification and validation of simulation models. In: Mitre systems engineering guide, pp 461–469. www.mitre.org/publications/technical-papers/the-mitre-systems-engineering-guide. Accessed 9 June 2015
8. Murray-Smith DJ (1995) Continuous system simulation. Chapman & Hall, London
9. Barton PL, Lee CK (2002) Modeling, simulation, sensitivity analysis and optimization of hybrid systems. ACM Trans Model Comput Simul (TOMACS) 12(4):256–289
10. Galán S, Feehery WF, Barton PI (1999) Parametric sensitivity functions for hybrid discrete/continuous systems. Appl Numer Math 31(1):17–47. doi:10.1016/S0168-9274(98)00125-1
11. Eker S, Singer JH, Yucel G (2011) Investigating an automated method for the sensitivity analysis of functions. In: Proceedings 29th international conference of the Systems Dynamics Society, Washington, DC

Chapter 5
Experimental Data for Model Validation

As mentioned in earlier chapters, uncertainties about the structure or parameters within a physically-based model can severely limit its applicability. In many engineering applications there are inherent difficulties in modelling the system dynamics on an entirely physical basis from first principles. Problems arising from uncertainties in the details of many aspects of the physical model contribute to these difficulties. There is therefore growing interest in the use of experimental techniques for establishing models from experimental data and from a combination of physically-based modelling principles and experimental methods.

5.1 An Introduction to Experimental Modelling Techniques

Many examples of the use of experimental modelling techniques may be found in the field of engineering control systems and in other areas of application. The case study of Chap. 9 describes the use of experimental modelling techniques in the development of a dynamic model of a laboratory scale process system involving two coupled tanks. Chapter 10 also provides useful illustrations of practical problems of simulation model validation in the context of helicopter flight mechanics modelling. Situations that necessitate the use of experimental modelling methods are also encountered in the modelling of physiological systems and in biomedical engineering applications involving the use of simulation models. The case studies of Chaps. 11 and 12 provide relevant examples of this kind in the context of physiological system applications.

System identification and parameter estimation are terms used, originally in the control engineering field and now more widely, to describe the processes used to derive models from sets of experimental data. The concept of system identification and the reason why it is a form of experimental modelling can be best understood

D.J. Murray-Smith, *Testing and Validation of Computer Simulation Models*,
Simulation Foundations, Methods and Applications,
DOI 10.1007/978-3-319-15099-4_5

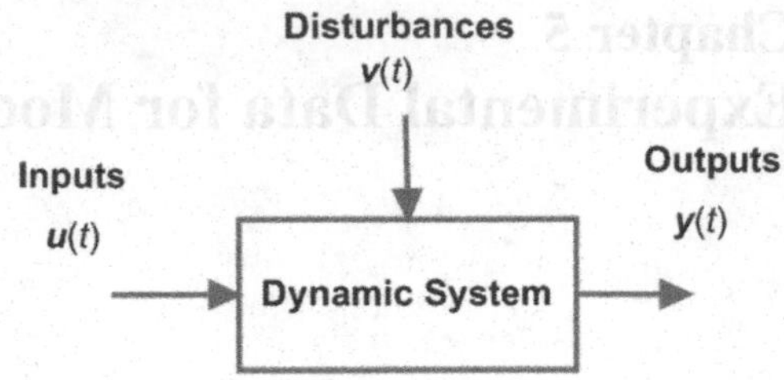

Fig. 5.1 Block diagram of a general dynamic system with inputs $\boldsymbol{u}(t)$ outputs $\boldsymbol{y}(t)$ and disturbances $\boldsymbol{v}(t)$

from Fig. 5.1 which shows a simple dynamic system with inputs $\boldsymbol{u}(t)$, outputs $\boldsymbol{y}(t)$ and disturbances $\boldsymbol{v}(t)$. The experimenter can control $\boldsymbol{u}(t)$ but not $\boldsymbol{v}(t)$ and the outputs $\boldsymbol{y}(t)$ (along with the inputs $\boldsymbol{u}(t)$) contain information about the system which can be used to fit a model to the available experimental data through an appropriate choice of model structure and parameters. System identification and parameter estimation is generally considered to be a mature field and classical methods of identification involve linear discrete-time or continuous-time models within a stochastic framework.

In general, system identification is a procedure that involves establishing possible model structures from experimental data as well as estimating values of parameters within that structure. In physically or biologically based models the structure of the model, which is often nonlinear in form, has been determined by the application of scientific laws and principles. In many cases the problem becomes one of parameter estimation, rather than the complete process of system identification. What we then have to do is to find the set of parameters which is optimum (or, at least, appropriate) and many different approaches may be used [1].

It should be noted that the process of adjusting parameters within a nonlinear computational model in order to improve the agreement with real-world information has been termed "calibration" by some. However, system identification techniques and the broader processes of model calibration should never be confused with model validation, which is entirely separate process although it also uses experimental techniques. Model calibration must be completed before any model validation processes are considered (see, e.g. [2]).

Together with issues of experimental design and the choice of test signals for the estimation of parameters, the selection of the model structure can contribute in an important way to the overall robustness of models that are calibrated using experimental data. The process of extracting data from system and sub-system tests is not a trivial task and the whole iterative process of model development in the presence of parametric or structural uncertainties raises many important issues. It also emphasises the fact that there are no generally accepted standard approaches to model evaluation. Subjective methods, including graphical and visualisation techniques are appropriate and are widely used as well as quantitative measures of various kinds, as outlined in Chap. 3.

The use of optimisation techniques within the model development process for the adjustment of model parameters has much in common with the use of optimisation in design and it is therefore helpful to apply experience gained in design applications to modelling situations. Although gradient-based optimisation methods remain important, the complexity of many practical problems means that it is impossible to establish a global optimum using gradient methods alone and more general techniques such as directed-search methods and evolutionary algorithms, can provide benefits. There is an obvious difficulty with this approach in that there is a serious danger of over-parameterisation where the number of parameters means that, although the techniques may produce a set of parameter estimates, these estimates may not lead to a satisfactory outcome when the model is tested using data sets that were not used for the estimation process. Models that include many unknown or ill-defined parameters that must be determined experimentally present particular difficulties in terms of parameter estimation as it is often possible to fit a very wide range of system responses if there are enough adjustable parameters in the model. It is said that John van Neumann stated "With four parameters I can fit an elephant and with five I can make him waggle his trunk" [3]. More recently, it has been shown [4] that, using complex quantities, the original statement about drawing an elephant-like shape with four parameters is actually correct! Tests of identifiability may help to overcome this problem of ill-defined parameters but it is always a potential danger with large and complex models.

As discussed in Chap. 4 another important topic, closely related to Parameter-optimisation, is parameter sensitivity analysis. Insight gained from parameter sensitivity information can be of considerable value in the development and refinement of system models through investigation of model robustness and the design of appropriate test inputs (see e.g. [5–7]).

5.2 Experimental Modelling Methods and Their Role in Model Development and Validation

As emphasised in Chap. 1, models must be appropriate for their intended purpose. Models are never unique and their development is an iterative process, involving initial formulation and testing, followed by repeated modification and re-testing. The form of model used at a particular stage in a project must therefore take account of the objectives, the amount of detail needed at that stage and uncertainties in the information about the real system. In some situations, particularly when working with existing systems or sub-systems, there is often a need for experimental investigations before a detailed physically-based quantitative model can be developed fully.

General issues arising in the validation of models or sub-models derived experimentally using system identification and parameter estimation techniques are considered further in Chap. 7 and receive further attention in the case-study chapters, especially Chaps. 9, 10, and 11.

5.2.1 An Overview of System Identification and Parameter Estimation Techniques

System identification and parameter estimation are well-established techniques involving use of measured response information from a real system to develop, directly, a mathematical or computer-based dynamic model. In this approach the model usually has a general form involving ordinary differential or difference equations and an associated set of parameters that have to be estimated. In general, the structure (as defined by the number of differential equations and any associated algebraic relationships) also involves uncertainties and the most appropriate form may have to be established using measured response data. Traditionally, identification techniques have been applied to the development of control systems, especially self-adaptive and self-tuning systems. However, system identification and parameter estimation techniques also provide analytical and computational tools that offer additional insight at various stages in the development of physically-based models.

In many practical applications models developed using system identification and parameter estimation techniques are linear in form. Techniques of system identification and parameter estimation for ordinary differential equation models are presented and reviewed in many textbooks, such as those by Ljung [7], Söderström and Stoica [8], Nelles [9], and Raol et al. [10].

Many approaches to identification and system parameter estimation involve optimisation of a specified cost function involving the difference, based on appropriate functions of the errors between model responses and the corresponding measured variables. These cost functions are closely linked to some of the measures of model quality outlined in Chap. 3. Decisions about the most appropriate structure for the model usually require background knowledge and physical understanding of the system, as well as examination of available experimental data and consideration of the intended application. Once an initial structure has been established and uncertainties have been critically assessed, the parameters of the model can be estimated, usually iteratively, through the use of a specific cost function and optimisation method such as least-squares minimisation. The iterative processes of parameter adjustment continue until the responses of the model match those of the system to some pre-defined level.

Many optimisation techniques and methods for the iterative solution of nonlinear equations have been developed which are applicable to the system identification problem. The textbook by Raol et al. [10] provides a useful review of least squares methods in the context of modelling, system identification and parameter estimation. The treatment of optimisation methods presented in that book establishes links between the properties of classical gradient-based optimisation techniques and widely used identification and parameter estimation methods, such as generalised least squares and nonlinear least squares methods and also techniques such as the equation error, output error and maximum likelihood methods. Further information about the use of some of these techniques in the context of model validation may be found in the case study in Chap. 10 where their use in helicopter applications is discussed and in Chap. 11 which deals with a physiological system application involving pulmonary gas exchange models.

Nonlinear optimisation methods, which are especially relevant for estimation of parameters of nonlinear models may be classified as "local" or "global". Although they may converge, local methods tend to become stuck at local minima or maxima and an extremum elsewhere in the parameter space may be neglected. Global nonlinear optimisation methods may overcome this difficulty and rely on the inclusion of random components that help the algorithm to avoid becoming trapped. Well known global optimisation techniques include "simulated annealing" (SA) methods and evolutionary algorithms such as the "Genetic Algorithm" (GA) [11].

It should be noted that most emphasis is placed in this chapter on methods of system identification that relate to parametric models. Simple pulse response or step response testing can provide much useful information about a system which can be valuable in forming a model, but the initial descriptions obtained experimentally are non-parametric in form and results from these methods may be difficult to interpret when output variables are corrupted by significant amounts of measurement noise. Non-parametric techniques that involve some form of averaging, such as those based on correlation methods or spectral analysis, may provide benefits in terms of sensitivity to measurement noise but also lead to non-parametric forms of system model represented by impulse, step or frequency response functions. Any description in parametric form has to be derived from these initial non-parametric representations. Such models are therefore obtained in a less direct fashion, but the non-parametric descriptions can themselves provide additional physical insight through, for example, examination of the ordinary coherence functions discussed in Chap. 3. Small coherence values can indicate that there are problems with measurement noise or that there are significant nonlinear effects. This is discussed further in Chap. 7 and again in the case studies in Chaps. 10 and 12.

In the case of partial differential equation (PDE) models, the literature on system identification and parameter estimation is relatively sparse and most of the currently available methods for estimating parameters necessitate the repeated solution of the PDE, which is inevitably slow due to the computationally intensive nature of the task. A different approach has been proposed by Bar et al. [12] who have developed a two-stage type of method where unknown PDEs are initially modelled using multivariate polynomials of sufficiently high order and a best fit procedure involving least squares minimisation. Parameters of the PDE could then be determined, through a further least-squares process, from the estimated functions [13]. Other published methods include a parameter cascading approach and a fully Bayesian type of treatment [14]. These two methods both involve representing the unknown system using a non-parametric function while using the PDE model to regularise the fit.

In the structural dynamics field theoretical finite element models present potential difficulties in that modelling errors and uncertainties can be hard to estimate. However, it is possible to compare a relatively small number of relevant dynamic properties for a given model with corresponding measured quantities and to reach conclusions about the suitability or otherwise of the theoretical model. Input and output measurements on a real structure can, in principle, be obtained with appropriate actuators and vibration sensors at selected locations within the structure. Parameter estimation techniques can then be applied, as in other applications, and these may lead, for example, to comparisons of system and model properties in terms of eigenvalues and mode shapes (see, e.g. [15]).

In recent years there has been considerable interest in model validation using a more direct frequency response function approach based on the Frequency Response Assurance Criterion (FRAC) discussed in Chap. 3 [16]. This is believed to have potential advantages over other approaches in that it makes direct use of measured frequency responses and thus captures the true damped response characteristics of the system, including nonlinear effects. Although formal computational procedures based on the FRAC have been developed for updating of finite element models, it should be noted that visual examination of measured and theoretical frequency response functions can still provide important insight. Methods based on frequency response functions have also been developed which allow nonlinear coefficients to be estimated together with an associated linear model [17]. The potential importance of methods for the development of accurate dynamic models of structures lies both in the need for improved models for structural design and for the use of models for the purposes of early damage detection.

Details of methods of system identification and parameter estimation are readily available elsewhere and are not repeated in this book. Each approach has specific strengths and weaknesses in the context of a specific type of application. Well-documented software for system identification and parameter estimation is also available from a variety of sources, including Matlab® toolboxes. (e.g. [18]).

5.2.2 Model Structure Optimisation

Model structure optimisation can be regarded as a procedure for the optimisation of model complexity since the number of separate equations and the number of adjustable parameters, which provide a crude measure of complexity, depend very much on the chosen structure. Also, with more parameters, a model becomes more flexible since the number of forms of behaviour that could be exhibited increases. A model that is too simple will not capture the behaviour of the system and will give poor predictions. In addition, if the data available for parameter estimation and subsequent testing of the model are inadequate, even a relatively complex model may perform badly. Thus, the complexity of a model must always be appropriate for the intended task.

The performance of an identified model should be assessed through the use of a "test" dataset that is not the same as any "training" dataset used for estimating its structure and parameters. This allows the capability of the identified model to predict system outputs for experimental situations that are not exactly the same as those used in the identification process to be assessed. This property of the identified model is termed "generalisation" and is closely linked to concepts of model validation.

The terms "under-fitting" and "over-fitting" are extensively used in model testing. If a simple model cannot match observed behaviour, under-fitting may have occurred and the model structure should be reconsidered. On the other hand, if a relatively complex model is used and the training appears satisfactory, but the generalisation is poor, over-fitting may have arisen due, possibly, to bias on the estimated values of model parameters associated with noise on the data used for identification, or to an inappropriate model structure.

How identification and parameter estimation techniques are used depends on the intended application. Different models and methods may be appropriate depending on the purpose of the model. For example, for many control system applications, linear forms of model may be used since control system design methods frequently require linearised plant models. However, nonlinear models are often essential in other application areas.

Procedures for the estimation of parameters of physically-based models are often based on methods that involve searching for a solution that optimises an appropriate cost function or fitness function. The process thus attempts to minimise, in some sense, the difference between the model responses and those of the real system, as outlined in the section above in the context of finite element structural dynamics problems. For a search-space with a small number of possible solutions it is possible to examine all of the solutions and find the best one in a relatively short time or, in the simplest cases, to use a trial and error approach to the optimisation problem. Exhaustive search methods are unlikely to be of great value in practical modelling situations and an outline of computer-based optimisation methods commonly used for this purpose may be found in [1]. That material is not repeated here other than to stress the importance of evolutionary computing methods in this context. Essentially, evolutionary algorithms perform a directed search through simulation of the evolution of successive generations of a population of individuals using operations that lead to a process of "survival of the fittest", as in natural selection. Although considered simplistic by biologists, evolutionary algorithms have been found to provide powerful adaptive search mechanisms that avoid problems that are often encountered with other optimisation methods when dealing with complex modelling problems.

One evolutionary computing method that has been applied successfully to the problems of nonlinear model structure identification involves "Genetic Programming" (GP) techniques [19]. This is an optimisation method that involves automatically selecting elements of the model structure from a database of possible components and combining them together to form a complete model.

GP techniques allow the optimisation of a tree structure or computer program rather than optimisation of numerical values of parameters. The tree structure has variable length and involves nodes (which can be "terminal" nodes, signifying an input or a constant, or "non-terminal" nodes which perform actions on signals). An example of one small tree structure is shown in Fig. 5.2. Here the three terminal

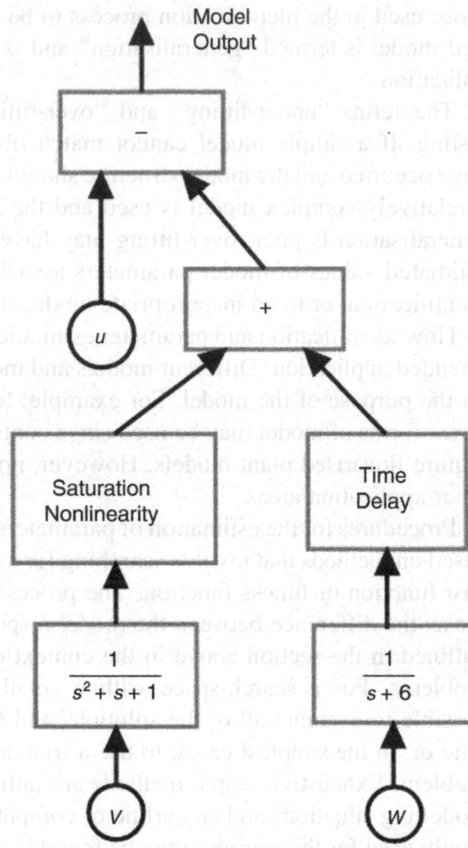

Fig. 5.2 Block diagram illustrating tree structure resulting from application of GP approach to model structure optimisation

nodes are associated with model inputs *u*, *v* and *w*. The remaining nodes (non-terminal nodes) are formed from standard model library blocks such as might be created within a block-oriented simulation tool such as Simulink® [20].

In order to carry out the identification in terms of a nonlinear model, the GP software selects candidate model blocks from within the model library to form a possible block diagram of the model. An appropriate "fitness function" is chosen to allow the quality of fit between measured response data sets and corresponding variables of the model to be assessed. By emulating natural evolutionary processes the GP attempts to generate a model structure that maximises or minimises the chosen fitness function. A randomly selected point is chosen and "crossover" and "mutation" operators act on the part of the tree structure below that point in the diagram. The crossover operation involves interchanging branches from two parent structures and mutation leads to creation of a random new branch. At each stage of the iterative optimisation process, known as a "generation", a population of possible model structures is subjected to processes of crossover, mutation and "selection". The chosen fitness function is then re-evaluated and the next generation is chosen from the old and new structures. The process is repeated then until some suitable convergence criterion is satisfied. A population of models thus evolves through many generations to produce a structure that is optimum in terms of some appropriate objective function. The set of parameters most suitable for the chosen structure is then generated in a second and separate stage of the system identification process. A parameter optimisation method which has been found to be particularly successful in this context involves a combination of simulated annealing and Nelder simplex methods [21].

It is very important to stress that in the GP approach, although the most appropriate structure is found in an automated process, knowledge and physical understanding of the system being modelled is very important. Such knowledge affects the choice of the functions to include in the library of components and the design of experiments to generate suitable data sets from the real system. This approach to nonlinear model structure identification is discussed further in the case study of Chap. 9.

5.2.3 *Issues of Identifiability*

The quality of a parameter estimate is normally expressed using the variance of the estimate. The variance depends both on the estimation method and on the experiment used for identification. In system identification the usual objective is to obtain unique and reliable estimates of all of the parameters of a model and it is important to know whether or not this is theoretically feasible for a specific form of model and particular experiment. The concept of "identifiability" is central to this and tests for identifiability allow potential problems to be found before an identification method is chosen and experimental design issues are considered.

"Global" or "structural" unidentifiability issues arise when a model has too many parameters to allow all of them to be found independently for any possible input stimulus. This depends on the form of the model equations (i.e. the model structure) and not on numerical values of the parameters or the design of the identification experiment.

This form of identifiability is only a minimal necessary condition for obtaining unique estimates of model parameters, as recognised by Bellman and Åström [22] who were among the first to discuss this issue. Their work on structural identifiability was carried out in the context of biological compartmental models, but the results may be applied to many other types of problem. They suggested that classical transfer function theory provides a basis for the investigation of identifiability. If each coefficient of the transfer function matrix is expressed as a combination of the unknown parameters, a set of nonlinear equations is defined. The model is then identifiable in a global or structural sense if these equations have a unique solution [22, 23]. This approach has been used successfully in a wide range of different types of application, including pharmacokinetics [24].

"Pathological" or "numerical" unidentifiability can arise if a structurally identifiable model is being used with experimental data sets that are inappropriate, due to the available record being too short in relation to system properties such as the dominant time constants or the period of oscillatory components of the response. It could also arise if the measured response data are inaccurate due, perhaps, to measurement noise. This issue was recognised as being important by Brown and Godfrey [25] who introduced the word "determinancy" in describing such situations.

Beck and Arnold [26] have proved that model parameters can be estimated only if the parameter sensitivity functions for the output variable with respect to each of the parameters are linearly independent over the range of observations. Numerical unidentifiability problems may be detected directly from the time histories of parameter sensitivity functions in simple cases. The problem can also be investigated in a more general and systematic way using properties of the sensitivity matrix:

$$X = \begin{bmatrix} \frac{\partial y_1}{\partial q_1} & \cdots & \frac{\partial y_n}{\partial q_p} \\ \vdots & \ddots & \vdots \\ \frac{\partial y_n}{\partial q_1} & \cdots & \frac{\partial y_n}{\partial q_p} \end{bmatrix} \tag{5.1}$$

and the closely related "parameter information" matrix $\boldsymbol{M} = \boldsymbol{X}^T\boldsymbol{X}$, where the variables y_i are the outputs of interest and the parameter sensitivities $\frac{\partial y_i}{\partial q_j}$ give an indication of the extent to which the variable y_i is influenced by a parameter q_j.

Analysis of this kind allows interdependencies to be investigated that are more complex than can be found readily by direct inspection of time histories of model variables or of the sensitivity functions. Pathological unidentifiability is associated with linear dependence of the columns of $\boldsymbol{X}$. This is reflected in the value of the determinant of matrix $\boldsymbol{M}$ or in the condition number of the matrix (the ratio of the largest eigenvalue of the matrix to the smallest eigenvalue). If the determinant is small, or if the condition number is large, the confidence region for the estimates will be large and the parameter estimates are unlikely to be accurate.

The inverse of the parameter information matrix ($\boldsymbol{M}^{-1}$) is also important and is known as the "dispersion" matrix, commonly denoted by $\boldsymbol{D}$. For an efficient estimator, such as a maximum-likelihood estimator, it may be stated that the elements of the dispersion matrix are, through the Cramer-Rao bound, related directly to the variance of the parameters found through system identification. The determinant of $\boldsymbol{D}$ can also be shown to be a useful indicator of numerical (pathological) unidentifiability [27]. It is also important to note that, in cases where the use of an efficient estimator is impractical, the dispersion matrix provides only a lower bound on the variance of estimates but test signals designed on the basis of the dispersion matrix may still be useful, although clearly not optimal.

Correlations between pairs of parameters can be detected using the "parameter correlation matrix", denoted by $\boldsymbol{P}$. This matrix may be defined in terms of its elements:

$$p_{ij} = \frac{m_{ij}^{-1}}{\sqrt{m_{ii}^{-1} m_{jj}^{-1}}} \tag{5.2}$$

where p_{ij} is the element of $\boldsymbol{P}$ in row i and column j and m_{ij}^{-1} is the element of $\boldsymbol{M}^{-1}$ in row i and column j. The diagonal elements of the matrix $\boldsymbol{P}$ have values which are unity and all the off-diagonal elements have values between -1 and 1. A model will be close to being unidentifiable if the modulus of one or more of the off-diagonal terms is close to unity and a value of 0.95 is regarded as a limiting value. On the other hand small values of the off-diagonal elements of $\boldsymbol{P}$ show that the parameters are essentially decoupled. This type of analysis is applied in Chap. 11 to the case study involving models of respiratory gas-exchange processes.

5.2.4 Design of Experiments and the Selection of Test-Input Signals

For the design of experiments and selection of test signals for system identification and parameter estimation it is assumed that the estimator is efficient and that these experimental aspects are independent of the estimation method. We must also have a quantitative basis for making comparisons of the effectiveness of different inputs.

Simple test inputs such as step inputs, pulses and sinusoids are widely used in experimental modelling. One of the problems with step and pulse inputs is that they are not "persistently exciting" as they involve significant time periods when they are constant or have zero value. Other difficulties with simple forms of test input is that their frequency components may not cover a sufficiently large part of the frequency range of importance for the system being modelled. In the case of traditional testing using sine-wave inputs it is necessary to apply different inputs, one after the other, with different frequencies, and the system must be allowed to reach a steady-state after each change of frequency. This can be very time consuming if a large number of different frequencies need to be applied in sequence to cover the frequency range of interest.

One approach that eliminates some of the problems that arise with these simple forms of test input involves the use of swept sine waves, also known as frequency-sweep inputs. This is a more complex from of signal in which the frequency is varied continuously from some initial low value to the uppermost frequency of interest. This is a persistently exciting signal which can be used to excite the system over a specified range of frequencies. Use of this test input greatly reduces the experiment time required compared with use of a sequence of conventional sinusoidal signals to obtain a measure frequency response. Difficulties can, however, be experienced with frequency-sweep signals in the case of systems that have significant oscillatory components since the test input may produce large and, possibly, nonlinear responses at frequencies close to frequencies of marginally stable or unstable modes. This may negate assumptions of linearity in the model being identified and may also mean that the response moves well away from the chosen operating condition.

Another approach to the design of persistently exciting signals can involve what are termed binary multi-step *i*nputs. As the name suggests, these have only two possible values and consist of sequences of step-like changes between those levels. Ideally, such signals should also have no component at zero frequency (i.e. no "dc" component, in electrical engineering terminology). Pseudo-random binary (PRB) signals are periodic binary multi-step signals of this kind which have an auto-spectrum similar in form to that of band-limited white noise. They can be used to excite a system over a specific range of frequencies and have proved very successful in the development of non-parametric models in the time domain and in the frequency domain. It should be noted that PRB signals do show dc offsets but the effect of this can be compensated for as it is small in the case of most signals used in practical applications.

Other forms of binary multi-step signal have been applied successfully in system identification and parameter estimation and also for model validation. As with all other forms of test signal, it important to have objective quantitative measures that make it possible to compare the effectiveness of different types of multi-step input. This can involve quantities such as the parameter information matrix and the dispersion matrix. As shown in Sect. 5.2.3, these matrices are based on the parameter sensitivity matrix $\boldsymbol{X}$, which can be found from measurements and the elements of the dispersion matrix can thus be derived from tests. In general, inputs giving small dispersion matrix elements are preferred over inputs producing large values.

The sensitivity matrix $\boldsymbol{X}$, the parameter information matrix $\boldsymbol{M}$ and the dispersion matrix $\boldsymbol{D}$ are therefore all used in establishing measures of the quality of experiments using relationships that can be described by the general expression:

$$J = f(\mathbf{M}) \tag{5.3}$$

where f is an appropriate scalar function.

There are a number of different performance measures that are based on the dispersion matrix, $\mathbf{D}$. These include an A-optimal criterion which involves minimisation of the trace of $\mathbf{D}$, an E-optimal criterion which involves minimisation of the maximum eigenvalue of the matrix $\mathbf{D}$ and the D-optimal criterion in which the determinant of $\mathbf{D}$ is minimised. The D-optimal criterion has a number of advantages over the other approaches since it is invariant for scale changes in parameters and for linear transformations of the associated models [27] and is probably the most widely used criterion for experiment design. It has the form:

$$J_D = \det(\mathbf{D}) = \det\left(\mathbf{M}^{-1}\right) \tag{5.4}$$

This criterion gives a test signal that places equal emphasis on the estimation of all of the parameters. If only a subset of parameters is of particular interest, it has been suggested that a "truncated" D-optimal criterion should be applied [28], and this has the form:

$$J_{Dt} = \det\left(\mathbf{M}_{ii}{}^{-1}\right) \tag{5.5}$$

where $\boldsymbol{M}_{ii}$ is a sub-matrix of the full information matrix involving only to the i parameters of interest.

These D-optimal criteria involve elements of the sensitivity matrix $\boldsymbol{X}$ which themselves depend on the model parameters. Thus it is only possible to use the criteria to investigate and compare experimental designs in a general way and it is not possible to generate a truly optimal experiment. To do that we would need a perfect model – which is what we are striving to find!

An interesting review of optimal experimental design issues in modelling of dynamic systems may be found in a paper by Titterington [29]. Although the paper is by no means recent it is still important in that it relates test signal design to more fundamental concepts of statistical optimal design theory and includes a comprehensive list of references to research on this topic.

There are many published accounts of applications in which identifiability analysis and experiment design have been used to good effect. Although much early research on identifiability analysis involved biomedical applications, the techniques have also been applied in other fields. Chapter 11 includes discussion of identifiability issues and questions of experiment design in connection with a nonlinear model of human pulmonary gas exchange processes.

One important class of signals used for linear system identification involves what are known as "multi-frequency" signals. These are periodic in form but have a spectrum which can be chosen with complete freedom and may therefore be matched to known properties of the system under test and any information about the spectrum of measurement noise. In the context of system identification the use of multi-frequency signals offers additional advantages over other forms of test signal in terms of flexibility in the choice of estimation method and also in the form of linear model. Both continuous system descriptions and discrete-time models can be derived. In another approach, described by van den Bos [30], time-domain data are transformed into Fourier estimates prior to parameter estimation. Benefits of this include the fact that the estimation process is formulated as a linear problem and also that the Fourier coefficients are easily interpreted. A broader discussion of test signals for frequency-domain system identification has been provided by Godfrey in [31] and a useful account of methods for the design of broadband excitation signals for the identification of parametric and non-parametric models has been provided by Schoukens et al. [32]. A useful paper describing the application of multi-frequency signals to the identification of a number of practical systems has been published by Harris [33]. The key issue is that, for the identification of a linear model, the perturbation must of sufficient amplitude to allow the response to be detectable in the presence of measurement noise but not so large that the system moves away significantly from the chosen operating point.

The issue of test signal design is returned to in Chap. 10 where problems of helicopter flight mechanics model validation are discussed. In that application the experimental design and choice of test signals is found to be of critical importance due to the presence of unstable modes. In that particular context both frequency sweep signals and binary multi-step test inputs designed in the frequency domain have been found to be of considerable value.

5.2.5 *Accuracy of Estimates*

One important issue in system identification and parameter estimation is the accuracy of parameter estimates. This also has important links to the form of model chosen since the reasons for finding a large scatter in parameter estimates often relates to issues of model structure.

In general, there are two broad categories of approach. The first of these may be termed scatter analysis and involves direct use of information about the statistical scatter of repeated parameter estimations. This approach can be applied when many sets of repeated test measurements are available for the same test conditions. Such a procedure is not usually appropriate for practical applications involving continuous system simulation models since the number of repeated estimates available is normally too small to provide results that are statistically significant.

The second approach to the assessment of the accuracy of estimated parameters is usually termed theoretical accuracy analysis. The underlying theoretical basis for this approach is provided by the Cramer-Rao Inequality which is used to define a Cramer-Rao bound which is always less than or equal to the standard deviation of the corresponding estimates of the parameter that would be found from scatter analysis of a large number of repeated tests. Poor identifiability for specific parameters is indicated by relatively large Cramer-Rao bounds and may suggest that such parameters should be fixed quantities within the model or should be removed from the model altogether. Although the Cramer-Rao bounds provide theoretical values for the standard deviations of parameter estimates it should be noted that, in commonly used methods of system identification, such as the maximum likelihood approach, factors of 5 or 10 are often introduced to provide a more realistic estimate of scatter. Such factors are needed to account for modelling errors and the effects of non-Gaussian noise, which can be reduced through appropriate filtering [34]. A factor of 2 is still recommended even for cases where the noise is properly modelled or eliminated. Tischler and Remple [35] suggest that the Cramer-Rao bounds are best expressed as a percentage of the parameter estimate. Thus, for a parameter θ_i

$$\overline{CR_i} = \left|\frac{CR_i}{\theta_i}\right| \times 100\,\% \tag{5.6}$$

A guideline suggested by Tischler and Remple [35] for a maximum acceptable value of $\overline{CR_i}$ for a parameter of a linear state-space model is 20 % for acceptable predictive accuracy but those researchers also suggest that, in practice, several of the Cramer-Rao bounds can exceed that value, even up to 40 % without serious consequences in terms of model quality.

References

1. Murray-Smith DJ (2012) Modelling and simulation of integrated systems in engineering: issues of methodology, quality, testing and application. Woodhead Publishing, Cambridge
2. Oberkampf WL (2007) Predictive capabilities in computational science and engineering. Presented at OASCR applied mathematics PI meeting, Lawrence Livermore National Laboratory, 22–24 May 2007. http://science.energy.gov/~/media/ascr/pdf/workshops-conferences/mathtalks/Oberkampf.pdf. Accessed 9 June 2015
3. Dyson F (2004) A meeting with Enrico Fermi. Nature 427:297
4. Mayer J, Khairy K, Howard J (2010) Drawing an elephant with four complex parameters. Am J Phys 78:648–649
5. Tomović R (1963) Sensitivity analysis of dynamic systems. McGraw-Hill, New York
6. Frank PM (1978) An introduction to system sensitivity theory. Academic, London
7. Ljung L (1999) System identification: theory for the user, 2nd edn. Prentice Hall, Upper Saddle River
8. Söderström T, Stoica P (1989) System identification. Prentice-Hall, New York
9. Nelles O (2001) Nonlinear system identification. Springer, Berlin
10. Raol JR, Girija G, Singh J (2004) Modelling and parameter estimation of dynamic systems, IET control engineering series no. 65. IET, London
11. Goldberg DE (1989) Genetic algorithms in search, optimization and machine learning. Addison Wesley, Boston, MA
12. Bar M, Hegger R, Kantz H (1999) Fitting differential equations to space-time dynamics. Phys Rev E 59:337–342
13. Müller TG, Timmer J (2002) Fitting parameters in partial differential equations from partially observed noisy data. Phys D 171:1–7
14. Xun X, Cao J, Mallick B et al (2013) Parameter estimation of partial differential equation models. J Am Stat Assoc 108(503). doi:10.1080/01621459.2013.794730
15. Ewins DJ (2000) Basics and state-of-the-art of modal testing. Sãdhanã 25(3):207–220
16. Grafe H (1998) Model updating of large structural dynamics models using measured response functions. PhD dissertation, Imperial College, London
17. Kershen G (2002) On the model validation in non-linear structural dynamics, doctoral dissertation, University of Liege, Belgium
18. Simscape™ software, Mathworks Inc., [Online] http://www.mathworks.com/products/. Accessed 5 June 2015
19. Koza JR (1992) Genetic programming: on the programming of computers by means of natural selection. MIT Press, Cambridge, MA
20. Matlab®/Simulink® modelling and simulation software, Mathworks Inc. http://www.mathworks.com/products/. Accessed 5 June 2015
21. Gray GJ, Li Y, Murray-Smith DJ et al (1996) Structural system identification using genetic programming and a block diagram oriented simulation tool. Electron Lett 32(16):1422–1424
22. Bellman R, Astrom KJ (1970) On structural identifiability. Math Biosci 7:329–339
23. Grewal MS, Glover K (1976) Identifiability of linear and nonlinear dynamical systems. IEEE Trans Autom Control 21:833–837

24. Milanese M, Molino GP (1975) Structural identifiability of compartmental models and pathophysiological information from the kinetics of drugs. Math Biosci 26:175–190
25. Brown F, Godfrey KR (1978) Problems of determinacy in compartmental modeling with application to bilirubin kinetics. Math Biosci 40:205–224
26. Beck JV, Arnold KJ (1977) Parameter estimation in science and engineering. Wiley, New York
27. Cobelli C, Romanin-Jacur G (1976) Controllability, observability and structural identifiability of multi input multi output biological compartmental systems. IEEE Trans Biomed Eng 23(2):93–100
28. Hunter WG, Hill WJ, Henson TL (1969) Designing experiments for precise estimation of all or some of the constants in a mechanistic model. Can J Chem Eng 47:76–80
29. Titterington M (1980) Aspects of optimal design in dynamic systems. Technometrics 22(3):287–299
30. van den Bos A (1993) Periodic test signals – properties and use. In: Godfrey KR (ed) Perturbation signals for system identification. Prentice-Hall, Hemel Hempstead, pp 161–175
31. Godfrey KR (ed) (1993) Perturbation signals for system identification. Prentice-Hall, Hemel Hempstead, pp 60–125
32. Schoukens J, Guillaume P, Pintelon R (1993) Design of broadband excitation signals. In: Godfrey KR (ed) Perturbation signals for system identification. Prentice-Hall, Hemel Hempstead, pp 126–160
33. Harris SL (1993) Generation and application of binary multi-frequency signals. In: Godfrey KR (ed) Perturbation signals for system identification. Prentice-Hall, Hemel Hempstead, pp 209–223
34. Maine R, Iliffe R (1985) Identification of dynamic systems – theory and implementation. NASA report RP-1138, NASA, Dryden Flight Research Center, Edwards, CA
35. Tischler MB, Remple RK (2006) Aircraft and rotorcraft system identification. AIAA, Reston, VA

Chapter 6
Methods of Model Verification

6.1 General Issues in Simulation Model Verification

As discussed in Chap. 2, the word "verification" describes the process of establishing that a computer simulation is consistent with the underlying conceptual or mathematical model. This usage of the word "verification" has also been adopted by many groups dealing with more specialised areas within the modelling and simulation field. For example, the American Institute of Aeronautics and Astronautics Committee (AIAA) on Standards in Computational Fluid Dynamics (CFD) has defined "verification" as the "process of determining that a model implementation accurately represents the developer's description of the model and the solution to the model" [1]. The procedure is internal to the simulation process and is not concerned with the suitability or otherwise of the underlying model in mathematical or conceptual form. It has therefore been suggested in the past that the word "internal" could be coupled with the word "verification" to distinguish this procedure from the process of validation [2]. The latter may then be termed "external validation" in recognition of the fact that information from the real world, external to the model, is being used. The processes of internal verification of a simulation model are similar to the processes used in testing of software more generally and some well-established software testing principles can be used (see e.g. [3, 4]).

The process of internal verification involves:

- As a first step, careful consideration of the structure of the simulation program and of the model upon which it is based to show that they are internally consistent and that there are no contradictions in terms of mathematics, logic or organisation This aspect of verification may thus be said to relate to the "computational model" and involves a process of code verification, including finding and removing errors in source code and also within numerical algorithms. This code verification stage of the process involves building on any

D.J. Murray-Smith, *Testing and Validation of Computer Simulation Models*,
Simulation Foundations, Methods and Applications,
DOI 10.1007/978-3-319-15099-4_6

previous software testing that has been carried out to establish the correctness and robustness of the code in the context of the application of the simulation program. There are three commonly used approaches to the problem of detecting run-time errors in software code arising in simulation models. These are code reviews, which involve checking every line of a program for potential errors, static analysis and trial-and-error dynamic testing. Code reviews are highly labour intensive and therefore may be difficult to apply for large and complex simulation models. Static analysis, as the words suggest, involves testing under steady-state conditions, for which results can often be found analytically. However, static analysis may leave much of the source code untested. Dynamic testing removes the restrictions that apply to the static case but requires a lot of different test cases to be devised, run and then interpreted carefully. When a test fails, additional time must be spent to find the cause of the problem. Examples of possible reasons for run-time errors include attempts to access an array outside the array bounds and sections of code that are unreachable because of programming mistakes. Code verification is the responsibility of both the code developer and the model developer, if they are not the same person. In many organisations, code verification is likely to involve the use of software quality assurance standards and procedures. These ensure that the code is reliable in the sense that it is implemented correctly for the specified computer hardware and operating systems etc. on which the simulation model will be used (see e.g. [3, 4]).

- The second stage of the internal verification process is concerned with demonstrating the accuracy of the data used for the simulation and estimating errors in numerical solutions. This aspect of the process relates to the "computational solution". The simulation model must be evaluated for special cases, each of which usually involves examination of a specific aspect of the simulation model to ensure that the model provides a sufficiently accurate solution. In the case of a model based on ordinary differential or differential algebraic equations these tests may involve investigation of the effectiveness of the chosen simulation algorithms in terms of integration accuracy and the accurate detection of discontinuities. Here the model developer plays an important part by trying to devise tests that involve comparison of highly accurate solutions for specific well-understood cases, when available, with solutions obtained using the simulation model software.

It is important to note that errors in the context of verification can be defined as deficiencies that are not due to lack of knowledge. These may be divided into acknowledged errors and unacknowledged errors. Acknowledged errors are errors that can be estimated or bounded in some way (such as those associated with a specific numerical algorithm) while unacknowledged errors are simply due to mistakes (in coding, for example).

Internal verification procedures are needed at every stage of the development of a simulation. Any change within a model must lead to further verification of the associated simulation program. At every stage there must be some form of verification procedure involving comparison between the computational solution and a reference solution provided in some other way from the underlying model, even if this has to be for very specific situations. For example, some cases may allow analytical solutions to be found and such a solution then forms a standard against which the numerical solution can be compared.

Some aspects of internal verification are closely linked to software quality assurance practices such as configuration management, software quality analysis and testing. Common-sense procedures that should always be applied, however simple the simulation model, include the following:

- in the case of simulation models developed using conventional programming tools, a graphical representation should also be developed for the model and be included in the documentation.
- a flow-diagram that includes all logically possible outcomes that can occur during system operation should be developed and should form part of the documentation.
- the model and the associated simulation program, together with all relevant documentation, such as graphical representations and flow diagrams, should be checked by someone who was not responsible for the development of the model.
- when carrying out test runs for a simulation program, values of parameters provided as input should be displayed at the end of the run to check that they have not been altered by mistake during the simulation.
- the user should consider creating an animation to be used to check that what is observed from the simulation model does not show any unexpected features, in terms of what is known about the behaviour of the real system. Such an animation is often much more understandable than multi-output records of model variables. Unexpected behaviour in one of these records, such as a transient numerical instability, might be missed in examining multi-channel graphical records but could be much more obvious in an animation.

Since there is never any proof that a given computer program is free from errors, the use of a number of well-designed test cases is critically important for verification. These test cases must be fully documented and must include information about the computer hardware, operating system and other software used.

Thus, traditionally, establishing whether or not a set of verification tests is sufficient for the intended application is in many ways more of an art than a science. However, some new approaches are being introduced. One example that provides a more objective measure of test completeness is Modified Condition/Decision Coverage (MC/DC) which has been widely used in avionic system development [5] "Coverage" is a measure of how much of the logic in a model has been exercised during testing and MC/DC is a stringent measure. For example, the U.S. Federal Aviation Administration's (FAA) DO-178C safety-critical standard [6] requires complete MC/DC of any software being deployed in a safety-critical system and model-based development and validation issues relating to this standard are also receiving attention [6]. This is a potentially important development, but it must be recognised that preparing a set of tests that achieve complete MC/DC is challenging.

6.1.1 Verification of Simulation Models Based on Ordinary Differential Equations and Differential Algebraic Equations

At the most basic level, code verification for continuous system simulation models based on ordinary differential equations involves line-by-line checks of the simulation program or of connections between block diagram elements if the simulation is developed using a graphical user-interface. Links with sub-programs must also be checked carefully if existing (and already accepted) sub-models are used. Checks should also be performed for special cases involving particular static or equilibrium conditions that can often be investigated from the underlying model using pencil-and-paper calculations.

Many different types of algorithmic errors can arise with continuous system simulation models. The overall accuracy of a simulation model may be very different from the formal order of accuracy for a single algorithm used within that model (e.g. for numerical integration).

A simulation model can involve numerical algorithms of many kinds and the links between different error sources can be subtle and complex. For example, truncation errors may arise that are due to the fact that perturbations in variables are not sufficiently small and do not lead to a solution within the asymptotic convergence region. There may also be singularities or discontinuities which, for some integration algorithms, can give rise to situations in which a hard nonlinearity may appear to be ignored within the solution. There can also be problems of quantisation due to round-off errors linked to the word length being used for variables within the simulation model.

All issues of this kind should become apparent during the internal verification process and, as always, need to be assessed carefully in the context of the intended application. The levels of uncertainty associated with parameters of the model and in the model structure also need to be kept in mind during this assessment. While it is important to make use of algorithms that are appropriate for the intended use of the model, there is little point in having excessive numerical accuracy if the basic uncertainties about the model structure and parameters are large.

Simple checks may also be made, usually for dynamic conditions, of the appropriateness of the user's options in terms of the selected integration method. For example, even if a suitable integration algorithm has been selected, use of an inappropriate integration step interval might result in numerical instability which could be interpreted, incorrectly, as a feature of the model rather than an artefact of the simulation program. Similarly, the communication interval used for plots of output variables and for control of data flow between the simulation and external hardware or the operator is very important. An incorrect communication interval could lead to some transients disappearing from the graphical output even if they were represented correctly in the simulation output. Thus simple tests involving changes of integration parameters and communication interval can often help to establish, quickly, the nature of any problem encountered during verification. For example, if small changes of integration step cause large changes in the overall behaviour of the model it is likely that the underlying problem is numerical and is a feature of the simulation program rather than indicating an error in the underlying model itself.

6.1.2 Verification of Models Based on Partial Differential Equations

Distributed parameter models, which are based on partial differential equations (PDEs) have to be discretised in terms of spatial dimensions as well as time and involve boundary conditions which may be quite complex in practical applications. Possible difficulties that can arise (in addition to coding errors) relate to errors in the discretisation process or a failure of the numerical solution to converge over the chosen grid.

In general, with distributed parameter models the quality of the solution is defined in terms of a truncation error and a discretisation error. The former is associated with the order of accuracy of the solution of the discretised equations while the latter is concerned with the errors in using a set of discretised equations to represent the original PDEs. Errors can also arise from the process of discretising boundary equations and auxiliary equations (which take the form of additional linear or nonlinear algebraic equations). Over-specification of discrete boundary conditions can lead to problems of numerical divergence while under-specification can lead to problems of non-convergence of the numerical solution. Numerical issues of this kind depend on the size of grid used [7]. Steinberg and Roache [8] present an interesting approach to verification using analytical solutions based on symbolic manipulation methods and this type of methodology has been further refined by Roache and presented in more recent publications (e.g. [9, 10]).

6.1.3 Verification of Discrete-Event and Hybrid Models

The verification processes for discrete-event models are, in many respects, very similar to those that arise with continuous system simulation models. They can again be divided into checks carried out to detect errors in coding and checks to ensure that the algorithms used are appropriate in terms of the intended application.

Coding checks depend, as before, on whether the simulation tools used are based on conventional programming languages or are based on a block-diagram oriented approach. The methods discussed above in connection with the verification of continuous system simulation models are still relevant. Algorithmic checks in the case of discrete-event models may be associated with some features that are not found so commonly in continuous system simulation models. Examples include checks that routines intended to produce specific random variables do in fact provide output having appropriate statistical properties.

In the case of hybrid models, which are part continuous and part discrete, the algorithmic checks must include features relating to continuous system simulation models, especially at interfaces between continuous elements of the model and discrete elements. For example, in a simulation model of a system representing a digital processor controlling some continuous plant, the checks must include careful assessment of the sub-models representing analogue-to digital and digital-to-analogue converters. Similarly, any hard nonlinearities in the model, such as saturation limits, static friction effects or backlash phenomena, require careful investigation at the verification stage to ensure that the timing of the onset of the nonlinear behaviour is sufficiently precise for the intended application.

6.2 The Role of Formal Methods in the Verification of Simulation Models

Formal methods for software quality assurance have been given some attention in the context of simulation model verification and validation in recent years. Clearly a formal proof of correctness is the most effective way in which a simulation model can be verified and validated. However, it is widely accepted that for, many practical applications, a formal proof of correctness is often unattainable for reasons of cost and time, and the challenge is now increasingly being recognised as being to determine when formal techniques can be applied in an effective fashion [11]. Although modelling and simulation applications have much in common with other types of computer application involving the need for high-integrity software, there are a number of features that are unique for the simulation case. The most obvious of these is often a need for simulation programs to be used to investigate phenomena beyond the operating regimes within which the model and system behaviour is well understood.

Current views of the use of formal methods for simulation models is that these techniques are best used as an aid in reasoning, rather than trying to obtain a "proof of correctness" which may be unachievable. For example, model checkers are available that can explore very large state spaces that cover, to some extent, all possible executions of a given program. They can demonstrate that a program has a desired property by examining all possible executions or can produce counter-examples where the property does not hold. Because the model is finite we know that the state-space search will terminate and this type of approach can provide a high degree of automation, at the expense of generality.

Code verification tools are now available which are based on formal methods. One example is the set of PolySpace® tools provided by MathWorks™ [12]. Such tools can analyse code and detect run-time errors of the types discussed above. Use of tools such as these can also provide important evidence about the absence of such errors and thus offers some assurance in terms of the reliability of the code.

References

1. Oberkampf WL (2007) Predictive capabilities in computational science and engineering. Presented at OASCR applied mathematics PI meeting, Lawrence Livermore National Laboratory, 22–24 May 2007. http://science.energy.gov/~/media/ascr/pdf/workshops-conferences/mathtalks/Oberkampf.pdf. Accessed 9 June 2015
2. Murray-Smith DJ (1998) Methods for the external validation of continuous system simulation models: a review. Math Comput Model Dyn Syst 4(1):5–31
3. Kit E (1995) Software testing in the real world. Addison Wesley, Harlow
4. Kaner C, Falk J, Nguyen HQ (1999) Testing computer software, 2nd edn. Wiley, New York
5. Heyhurst KL, Veerhusen DS, Chilenski JL, Rierson LK (2001) A practical tutorial on modified condition/decision coverage, NASA/TM-2001-210876. National Aeronautics and Space Administration, Langley Research Center, Hampton
6. Anonymous (2011) DO-331 Model-based development and verification supplement to DO-178C and DO-278A. RTCA, Washington, DC
7. Oberkampf WL, Blottner FG (1998) Issues in computational fluid dynamics code verification and validation. AIAA J 36(5):687–695
8. Steinberg S, Roache PJ (1985) Symbolic manipulation and computational fluid dynamics. J Comp Phys 57:251–284
9. Roache PJ (1997) Quantification of uncertainty in computational fluid dynamics. Annu Rev Fluid Mech 29:123–160
10. Roache PJ (1998) Verification of codes and calculations. AIAA J 56(5):696–702
11. Kuhn DR, Chandramouli R, Butler RW (2002) Cost effective use of formal methods in verification and validation. Invited paper, Presented at foundations '02 workshop, US Department of Defense, Laurel, Maryland, 22–23 October 2002
12. PolySpace® tools by MathWorks™. http://uk.mathworks.com/products/polyspace/?s_tid=brdcrb. Accessed 5 June 2015

Chapter 7
Methods for the Invalidation/Validation of Simulation Models

The testing process in the validation of simulation models is concerned with establishing the extent to which a model is an accurate representation of the real world from the perspective of the users of the model [1]. Compared with internal verification processes, external validation is a much more open-ended task that involves comparisons between the model behaviour and, ideally, the behaviour of the real system for the same chosen conditions. Some would claim that the word "validation" should never be used in the context of modelling since one can never fully validate a model. The emphasis should instead always be on "invalidation", since to validate a model would require an infinite amount of experimental data (see, e.g., [2]). Any claim that a model is valid may be refuted at any time when someone undertakes a different test yielding data that cannot be explained completely by the model. Although, within this book, the word "validation" is used for convenience in discussing this aspect of the processes of model testing, acceptance and accreditation, it must recognised that we are really discussing techniques of "invalidation".

7.1 An Introduction to Methods of Model (In)Validation

Most traditional approaches to validation are based on repeated use of simulation methods, but an exhaustive procedure in which every possible case is considered is clearly impossible. The required number of simulation runs to provide evidence of complete consistency would be very large and a "proof" of that kind can never be exact. Most techniques are more selective and involve a relatively small number of simulation runs together with the appropriate use of insight about the behaviour and properties of the real system to guide the testing process. Dynamic testing of an informal kind can be carried out only over a small fraction of the operating

D.J. Murray-Smith, *Testing and Validation of Computer Simulation Models*,
Simulation Foundations, Methods and Applications,
DOI 10.1007/978-3-319-15099-4_7

envelope. Quantitative and qualitative methods are commonly used but in most practical applications a combination of methods is applied.

Although it has been suggested frequently over the last two decades that "formal methods" have a potential role in external validation, there is little evidence at present that formal methods are being used within the validation processes for serious modelling problems [3]. There are at least three difficulties. One problem is that the ability of a model to make accurate predictions, within the context of its intended application, is a necessary but not sufficient condition for validity. Another obstacle to the adoption of formal methods for validation is that, at present, their use requires a substantial investment in terms of additional time and effort within the model development process and it is believed that this is likely be seen as affordable only in some specific safety-critical applications [3]. However, the fact that these methods are now having an influence within the semiconductor industry where companies that develop microprocessor chips and other complex electronic hardware are beginning to use tools based on formal methods to detect flaws in design, suggests that it is likely to be only a matter of time before these techniques become more widely used in simulation model development. It is possible that one of the first areas where formal methods will have an impact is in terms of model requirements. It is known that ambiguous, incomplete and inconsistent sets of requirements specifications in large software projects may lead to defects that can be very costly to correct. It is now recognised that formal specification languages can be used to represent sets of requirements and that these can be use to check for properties such as completeness and consistency using formal analysis techniques [4]. Once they are used routinely in the software development field, it is likely only to be a matter of time before such tools start being used within the simulation model development cycle.

7.2 Quantitative Methods

Quantitative methods of validation can be categorised in a various ways. Here, although there may be some overlap between different methods, quantitative approaches have been divided into five categories. These are simple predictive methods, methods based on system identification and parameter estimation, methods based on barrier certificates, model distortion methods and methods based on parameter sensitivity analysis.

7.2.1 Simple Methods for Predictive Validation

The approach to quantitative validation most commonly discussed in introductory texts on computer-based modelling and simulation may be termed "predictive" validation. In this approach the model is used to predict the behaviour of the system and then this prediction is compared with the behaviour of the real system to find out how close they are, either graphically or using deterministic measures of the type discussed in Sect. 3.3. Whatever the chosen measures, the level of agreement is assessed by quantifying the difference between data from appropriately designed experiments and model outputs for the same experimental conditions. However, it is important to note that validation procedures should not be viewed simply as involving a process of comparison in which the experimental data sets are taken as being correct. Measurements also have their errors and uncertainties and these must be assessed carefully as part of the validation process. Taken together with uncertainties in the model this means that agreement may be best expressed in statistical terms using an estimated error and associated confidence limits. However, obtaining the information needed for statistical analysis of this kind may require considerable effort and expense in acquiring and analysing the data.

Understanding of the structure of the real system can be very helpful in simplifying the process of predictive validation. Essentially, one carries out the comparison between selected pairs of variables of the real system and model, while using other measured quantities from the real system as measured input variables of the model. This process allows parts of the model involving the most significant uncertainties to be isolated in a simple way. The validation of the two tank system in the case study in Chap. 9 provides a useful illustration of how this physical approach can help to simplify the problems of predictive validation.

Examination of the form of the residual time history formed from the difference between a model output and the corresponding output measured from the real system can be helpful. In particular, examination of the autocorrelation function of the residual time history can be revealing in terms of structural inadequacies within a model. Ideally, residuals should have properties which closely approximate those of white noise and this can be tested in practice through calculating the sample auto-covariance for a number of different lag values. If the covariance is found to be close to zero for a range of lag values, the residuals may be assumed uncorrelated and thus to approximate closely to white noise. However, if the model order is inappropriate the residual sequence will be correlated [5].

Another way of considering issues of model structure when experimental data are available from a real system involves considering how well the model structure can represent the experimental results for all combinations of possible parameters, or equivalently, how bad the best model may be for a given data set. If the minimum value of the chosen error measure for that model structure is too large, the model may not be capable of representing the experimental observations and a new structure should be considered. One relatively recent approach involves the concept of a "barrier certificate" [6], which is an interesting approach through which inconsistency between model and experimental data can be proved without the need for repeated simulation runs. This is discussed in more detail in Sect. 7.2.3.

7.2.2 *Methods Involving System Identification and Parameter Estimation Techniques*

The techniques of system identification and parameter estimation outlined in Chap. 5 provide a second approach that can be very valuable when dealing with a model of an existing real system, or in a design situation where a prototype system or test rig is available for experimental testing.

Whatever approach to model evaluation is used in a specific application, one must distinguish carefully between the processes of system identification and parameter estimation applied in the initial stages of model development, the tuning procedures used in model optimisation and the processes applied to establish the quality of the optimised model. The term "model calibration" may be used to describe the iterative procedures of optimisation and tuning used during model development.

As already emphasised in Chap. 5, model calibration is not the same as model validation as these processes take place at different points within the iterative cycle of model development. In Sect. 5.2.2 it is stated that an identified model should always be assessed using a "test" data set that is not the same as the data set that was used in estimating the structure and parameters of that model. This principle of using validation data that is different from the calibration data is of central importance. In the context of identification this step in the development of the model is often termed "generalisation" and relates to the capability of the model to predict system outputs for experimental situations that are not exactly the same as those used in the identification process.

If parameter estimation can be carried out successfully and checks using different forms of test input and different measured response data sets all suggest that the identified models give good predictions, confidence in the chosen model structure increases. If, as is likely, the parameter estimation process does not lead, immediately, to models which have good predictive capabilities, more insight can be obtained about the suitability of the model structure in the following ways:

(a) Any difference between an estimated parameter value and the corresponding theoretical value needs careful examination if the quantity concerned happens to be one which is well known and well understood (e.g. a parameter representing a well-known quantity, such as the acceleration due to gravity, g).
(b) As discussed in Chap. 5, if the standard deviation of parameter estimates is large for a number of sets of test data there may be problems with the model structure or with errors in the measured data sets used for the parameter estimation process.
(c) If time histories generated from differences between predicted variables from the identified simulation model and the corresponding measured variables from the real system show distinctive features that are common to different sets of records there is a strong possibility that the model structure is wrong.

Such features can often be seen most clearly if autocorrelation functions are computed for such difference signals.

(d) One possible difficulty relates to parameter correlation problems where two or more estimated parameters within a model have the same effect on a selected measure of performance. Identifiability analysis can often provide useful insight in such cases. Also, in dealing with a situation involving two correlated parameters, one way of moving forward is to fix one parameter at a value that is considered reasonable and then repeat the parameter estimation process to establish a new estimate for the other parameter. Further insight may then be gained by repeating the estimation process with the fixed and estimated parameters interchanged. Similar processes can be applied for cases involving more than two coupled parameters, but all situations involving parameter correlation, regardless of the number of parameters involved, require careful assessment before firm conclusions can be reached. In some cases problems of this kind may necessitate reconsideration of the chosen model structure.

Most currently-available techniques for system identification and parameter estimation relate to models which are linear. For a system with significant nonlinearities, linear system identification methods can provide useful insight through establishing linear models for different operating points and test signal amplitudes across the operating envelope of the system. The trends in terms of the values of key parameters of the identified models can then be compared with trends in the values of parameters of linearised descriptions derived from the nonlinear model for the same operating conditions and differences can provide useful insight. Similarly, comparisons of trends in these parameter sets as operating conditions are changed are important indicators of the performance of the nonlinear model and may lead to its credibility being strengthened or questioned. An example of this approach is given in Chap. 10 in the case study concerning helicopter modelling.

In the case of multi-input multi-output systems linear models are often derived using time-domain or frequency-domain methods of system identification, leading to a matrix of transfer functions. In such situations comparisons of a measured frequency response function with a model frequency response function for each of the input-output combinations may be helpful. Decisions about model adequacy may be made in each case, either using graphical methods or through use of a deterministic measure of goodness-of-fit, such as those discussed in Chap. 3. Calculation of the ordinary coherence functions (see Sect. 3.3.2) for each of the input-output pairs provides additional evidence about the significance of the transfer functions found and may permit some transfer functions to be ignored while others are retained within the multi-input multi-output model. Transfer functions that are eliminated from the model are those that show low levels of coherence over the complete frequency range of interest. In the context of helicopter flight mechanics modelling, Tischler and Remple [7] suggest that a sub-model with coherence values below 0.4 should be removed from a multi-input multi-output (MIMO) description. An alternative approach is to define an average overall cost function which is the sum of the individual cost functions for the transfer functions retained after coherence analysis, divided by the number of transfer functions involved.

If the level of agreement between the identified and theoretical models is considered adequate, a second stage of the external validation process can be attempted. This involves comparison of responses of the nonlinear model with the responses of the real system for larger perturbations and is based on direct comparisons of model predictions with measured system responses for the same variables. In some cases the validation may be attempted directly using simple graphical comparisons and methods involving quantitative measures. In such an approach, the polygon type of graphical display discussed in Sect. 3.5.1 has proved useful. If, once again, the level of agreement is judged to be acceptable over an appropriate range of conditions the model can be considered for release for the intended application. It can continue to be used until additional information or data give cause for concern in terms of quality.

When large inputs are applied to models having significant nonlinearities, traditional methods of validation based on direct comparisons of models and system have been found to have practical limitations, whether based on graphical methods or quantitative measures. More holistic methods involving the opinions of experts involved with the real system (e.g. pilots in the case of aircraft or operators in the case of industrial systems) may provide valuable insight concerning model limitations in such cases.

One approach that is well suited to the system identification or validation of physically-based continuous system models has been presented by Knudsen (e.g., [8]). This is based on parameter sensitivity information which is used in selecting model structures, for experimental design and for the validation of models identified from experimental data. The essential point of incorporating parameter sensitivity information into the identification approach is that accurate estimation of any parameter requires that the cost function upon which the estimation is based should be sensitive to that parameter and the most sensitive parameters are likely to be the ones that are estimated most accurately. Comprehensive knowledge of parameter sensitivities thus provides important information for assessing identification results.

It should be noted that non-parametric approaches to system identification, such as frequency-domain identification techniques, may be used to obtain empirical models to which physically-based descriptions may be fitted. For example, it may be possible to obtain transfer function parameter estimates using system identification methods and from these obtain parameters for a physically-based model, together (ideally) with error bounds. From these transfer functions one may then be able to derive values for physically based-parameters and find associated error-bound values. Comparisons with expectations from theory can then be made and sensitivity issues explored.

For linear single-input single-output models there are many methods for fitting transfer function descriptions to time-domain or frequency-response data obtained from experimental measurements. In the case of frequency-domain methods these measurements may, in some cases, be obtained using conventional testing methods based on the use of sinusoidal test signals with direct measurements of amplitude and phase shift of the output relative to the input. They may also be obtained using spectral analysis methods and appropriate broad-band test signals, such as pseudo-random binary (PRB) test signals or frequency-sweep inputs. However, it is important to note that linear transfer function models, obtained by frequency-domain testing methods, should be subjected to further assessment using test input signals that are different from those used for identification.

Examples of a procedure of this kind may be found in the case study on helicopter flight mechanics model validation, where parameter values within relatively low-order linear models, found experimentally using system identification methods, are compared with theoretical values for a range of flight conditions and conclusions are reached about the quality of the underlying nonlinear model.

This approach has also been applied to allow comparisons to be made, in the frequency domain, of reduced-order models with higher-order equivalents. This has been extended to the multi-input multi-output (MIMO) case and a method has been developed for ensuring that denominators of the transfer functions within a MIMO model, which in theory should be identical, are correctly estimated [9].

Using system identification and parameter estimation techniques often leads to the use of statistical techniques that are termed "frequentist" methods. Such procedures [11] are typically based on tests of hypotheses and allow a particular set of model parameters to be accepted or rejected. Early work by Balci and Sargent [10] was based on the multivariate Gaussian Hotelling T^2 test and more recent research on significance tests by McFarland and Mahadevan [11] has extended this, as has the work of Huynh, Knezevic and Patera [12].

Since validation can involve consideration of many different sets of parameters and boundary conditions, repeated solution of the model equations is inevitable and, in any application involving large and complex models, this can be time-consuming. In the case of models based on PDEs, repeated simulation runs (e.g. using finite element methods) can be computationally very expensive. Good management of the validation process is therefore very important to ensure that the maximum benefits are obtained from each simulation run and that potentially costly human errors are minimised. Model management issues are discussed in more detail in Chap. 8.

It should be noted that, in the context of biological models involving networks of neurons and continuous elements (such as muscle or sensory receptors), there are apparent difficulties in applying experimental modelling techniques due to the fact that some quantities within the model can be regarded as continuous in form (such as muscle length or tension) while others are inherently discrete, such as nerve signals which involve information transfer through the times of occurrence of short pulses known as action potentials. Some models of systems of this kind have been developed using appropriate methods for converting the discrete pulse signals derived from action potentials to equivalent quasi-continuous representation involving, for example, instantaneous frequency measures. However, a more powerful approach involves using point-process descriptions for the neural pathways and employing identification techniques that exploit similarities between identification methods for point-process systems and conventional identification methods for continuous systems [13].

7.2.3 Barrier Certificate Methods

The barrier certificate approach to model validation has been creating much interest in recent years. In this approach it is assumed that we have developed a nonlinear dynamic model

$$\dot{x} = f(x, p, t) \tag{7.1}$$

where $x(t)$ defines the state of the model and p is the set of model parameters. A set of measurements X, obtained from the real system over a time interval 0 to T, is also available. For the given model and a given parameter set and trajectory information X, we can invalidate the model if we can prove that, for all possible parameters, the model cannot produce a trajectory that follows the measurement trajectory exactly over the given time interval. This is done through attempting to create a barrier "certificate". If a barrier certificate is found to exist, it can be viewed as proving that there is a separation between measurement data and the corresponding trajectories obtained using the simulation model. Using this approach we can prove, in principle, that the model and its parameter set are inconsistent with the measured trajectory in state space without running any simulation. The major advantage of this approach over the use of repeated simulation runs for selected sets of model parameters is that the barrier certificate method involves a proof of invalidity which is exact. This can never be the case with traditional methods of model testing [6].

The basic theorem of barrier certificates requires the construction of a function that satisfies certain non-negativity conditions and this not an easy task in the case of models of practical significance. It has been likened to the task of finding Lyapunov functions in proving stability within nonlinear dynamic systems. In some cases, sum-of-squares decomposition techniques can be applied and solutions can be found using semi-definite programming methods for which a software tool (SOSTOOLS) is available. Anderson et al. [2] have published an interesting and easily accessed account of the application of the barrier certificate approach to a continuous time biochemical system model and also to a discrete-time model of population growth.

7.2.4 Methods Based on Model Distortion

The model "distortion" approach to simulation model validation has strong links with the techniques of system identification and parameter estimation. It involves determination of the "distortion" of parameter values needed to obtain outputs from the simulation model that match a set of measured data at each time point. The difference between the nominal values of parameters and the altered values then provides a measure of model quality. In general terms it can then be said that the smaller the amount of distortion required to achieve a match the more credible is the model.

The method, as presented by Butterfield et al. [14], applies to a nonlinear model described by ordinary differential equations in state space form:

$$\dot{\boldsymbol{x}} = \boldsymbol{f}(\boldsymbol{x}, \boldsymbol{u}, \boldsymbol{\beta}) \tag{7.2}$$

$$\boldsymbol{y} = \boldsymbol{C}\boldsymbol{x} \tag{7.3}$$

where $\boldsymbol{x}$ is a vector of n state variables, $\boldsymbol{y}$ is a vector of r output variables and $\boldsymbol{u}$ is a vector involving m input variables. The vector $\boldsymbol{\beta}$ is a p-vector of parameters. The matrix $\boldsymbol{C}$ in (7.3) is a full rank $r \times n$ matrix which has zero elements apart from one element in each row which has a value of unity.

Measured values at times t_j are given by a sequence of vectors z_j of q observations and these are assumed to be the noise-free responses of a system to known inputs $\boldsymbol{u}_j$ and $\boldsymbol{y}_j$ in the corresponding model response [15].

We now define a nominal model:

$$\dot{\boldsymbol{x}} = \boldsymbol{f}(\boldsymbol{x}, \boldsymbol{u}, \boldsymbol{b}) \tag{7.4}$$

$$\boldsymbol{y} = \boldsymbol{C}\boldsymbol{x} \tag{7.5}$$

in which $\boldsymbol{b}$ is an estimate of the parameters $\boldsymbol{\beta}$.

The model may be distorted so that:

$$\dot{\boldsymbol{\chi}} = \boldsymbol{f}(\boldsymbol{\chi}, \boldsymbol{u}, \boldsymbol{b} + \delta \boldsymbol{b}_j) \tag{7.6}$$

$$\boldsymbol{z}_j = \boldsymbol{C}\boldsymbol{\chi}\left(t_j\right) \tag{7.7}$$

and here $\boldsymbol{\chi}$ is the distorted state trajectory and $\delta \boldsymbol{b}_i$ are the model distortions needed to ensure that the outputs are the same at the times t_j. Since it is normally impossible to find a unique set of distortions the problem may be reformulated using a cost function:

$$J = \sum \delta \boldsymbol{b}_j^T \boldsymbol{W}_j \delta \boldsymbol{b}_j + \left(\boldsymbol{\chi}_j - \boldsymbol{x}_j\right)^T \boldsymbol{V}_j \left(\boldsymbol{\chi}_j - \boldsymbol{x}_j\right) + \lambda^T \left(\boldsymbol{z}_j - \boldsymbol{C}\boldsymbol{\chi}_j\right) \tag{7.8}$$

where $\boldsymbol{W}_j$ and $\boldsymbol{V}_j$ are weighting matrices. The third term on the right hand side of (7.8) involves a vector Lagrange multiplier $\boldsymbol{\lambda}$ and minimisation of the cost function provides a solution which corresponds to the smallest distortions which satisfy the requirement.

The time series of distortion values $\delta \boldsymbol{b}_j^T$ can be examined to see how they compare with any information that may be available, such as parametric uncertainties associated with component tolerances. Examination of the dependence of the cost function J on each of the parameters can also provide useful information [15].

Clearly, as with most other methods of model validation, errors and uncertainties in measured response data can present problems with this approach. However, as has been shown by Cameron et al. [15], the methodology can be applied usefully in considering the quality of linearised models for specific operating points, compared with the nonlinear description from which they were derived.

It is important to note that the distortion approach can be used both with single-input single-output (SISO) models and with multi-input multi-output (MIMO) descriptions and that the computational costs need not be excessive [15]. Numerical problems can arise if inputs and outputs fall to zero, as pointed out by Cameron et al. [15]. Although discussed by Thomas [16], by Cameron et al. [15, 17] and by Gray et al. [18], relatively few published accounts have appeared involving significant applications other than work reported by Butterfield and his colleagues in connection with problems arising in the nuclear industry [19–21].

7.2.5 Methods Based on Parameter Sensitivity Analysis

Although it has to be accepted that the optimal values for some parameters of a simulation model can never be known, it is usually possible to define intervals within which parameter values are expected to lie. This may involve use of knowledge of the real system or may involve repeated application of a variety of experiments designed for system identification and parameter estimation purposes. In general, it can be said that the quality of a model is in doubt if variation of a parameter within its expected interval leads to significant changes in the simulation results. Unless there is clear evidence, from the real system, that the overall behaviour is very strongly influenced by the physical equivalent of the parameter in question, a situation of this kind suggests that further investigations are needed. This may even involve tests on the real system in which the system itself is modified so that the equivalent quantity in the model is altered in a specific and systematic fashion (e.g. a mass is increased or decreased, or the position of a centre of gravity is changed, or a damping factor is changed, or a pipe diameter is changed). The techniques of parameter sensitivity analysis presented in Chap. 4 provide appropriate methods for model analysis in attempting to deal with situations of this kind [22].

It is also important to note that the individual effects of different model parameters may offset each other and this can cause difficulties in model validation unless the investigator is fully aware of the complexities of the situation. Parameter sensitivities are often not independent of each other and it is clearly impossible to consider any kind of systematic investigation of such effects for all of the parameters of a model using repeated simulation runs. At best we might consider investigation of cross-sensitivity effects of this type for a relatively small number of parameters that are believed to be particularly important.

Although the techniques of parameter sensitivity analysis outlined in Chap. 4 relate to the assessment of the effects of relatively small changes of model parameters on specific input-output responses in the time-domain (or in the frequency-domain), there is another aspect that needs to be considered in the context of model validation. This is the fact that defining a quantity within a model as a parameter rather than as a variable is a choice made by the model developer. For example, in a simple model of a dc electrical generator system, the electrical load might be represented parametrically as a resistor connected to the output of the generator. Sensitivity analysis could then provide information about the dependence of the voltage at the load on the load resistance value and such information could be helpful in assessing the effect of small variations about the nominal value. This could be an adequate model for many purposes, but there could be situations in which the electrical load on the generator would vary in a prescribed fashion over a period of time and the simple parametric representation in terms of a resistor would be incorrect. It would be appropriate to reconsider the model structure and introduce an additional physically-based sub-model to represent the load more accurately and allow quantities within that sub-model to vary with time.

In applications within the social sciences, where experiments on the real system for the purposes of model validation are generally impractical and also undesirable, there is an accepted need for some other kind of validation when models are being used for predictive purposes. In such situations, data based on historical observations may sometimes be of some value but there is usually considerable uncertainty about the precise conditions applying when such data were collected. Inevitably, this necessitates use of face validation methods but it has also been suggested that sensitivity analysis methods also may be of particular value for such cases. Since we are usually dealing with discrete-event models in social science applications the sensitivity calculations are most often approached using repeated simulation runs and differencing techniques [23].

7.3 Face Validation

An alternative to detailed quantitative comparisons of the model's performance with the response of the real system is a more holistic, but inevitably more subjective, assessment made by someone who has a deep and thorough practical understanding of the real system. This subjective approach is termed "face" validation and can be very helpful in establishing whether or not the logic within the conceptual model is correct and whether the input-output relationships for the model appear reasonable. This can be especially helpful in the early stages of engineering design and development projects when there is no prototype system available for testing. At a later stage of model development, after some quantitative validation has been carried out, it is sometimes possible to reintroduce an element of face validation with system experts being asked if they can discriminate between measured system and simulation output records. This is a form of Turing test and its main value is often to provide the model and simulation development team with insight regarding the way in which the experts approach this task and their reasons for reaching their decisions.

Interesting illustrations of the use of face validation may be found in some accounts of the development of simulation models used for real-time simulation applications in conjunction with external hardware to form a "hardware-in-the-loop" simulation. A common example of this is the development of control system hardware to be used, eventually, within a real engineering system. As a first step a real-time simulation model of the system to be controlled (the "plant") is developed and tested. If shown to be of acceptable quality, the real-time simulation may then be used as a test-bed in the design and development of the controller. Testing of the plant model is clearly of critical importance and many approaches are possible. Face validation is one approach commonly used, often in conjunction with other methods. People can usually be found who have expert knowledge of the type of system in question and often a wealth of experience in operations under normal conditions and for situations where faults occur. This is relevant for many different application areas, from flight simulators to simulators used in medical education or process control. The face validation approach is particularly important in checking the fidelity of a simulation in relatively uncommon fault situations.

One example of a project where face validation proved to be important, and was used alongside other validation methods, involved the development of a simulation model of a specific hydro-turbine generator system. This model involved not only the hydraulic, mechanical and electrical sub-systems of the turbine, generator and distribution network but also the pipeline system. The objective in developing the model was to provide a test-bed for evaluation of a range of different types of fast-acting electronic governors for speed control of the turbine and generator system. The specific plant modelled was at Sloy Power Station which was owned and operated by the North of Scotland Hydro Electric Board (NSHEB) at the time of the project [24].

A highly simplified schematic diagram of the system is shown in Fig. 7.1 and includes the reservoir and dam, a section of tunnel and pipe leading to a surge-shaft, before a lengthy down-hill section of pipeline providing the water supply to the turbine itself. Although the diagram shows only one turbine and one pipe the power station has four turbines and associated generators, operating in parallel, with each being supplied with water through its own final pipeline. Figure 7.2 is a slightly more detailed diagram of the pipe system and this shows that a single large diameter pipe divides, at a point below the surge shaft, into two pipes of smaller diameter which divide, in turn, into four pipes leading to the turbines. Only one of the four turbines and generators was assumed to be in use for this study and all modelling work and associated on-site testing was carried out on Turbine No 3. As show in Fig. 7.2, the pipelines leading to the other three turbines were closed at the turbine inlets. The rating of the turbine and generator under investigation was 35.75 MW and the electrical output could either be coupled to the national electrical power distribution network (the National Grid) or could be switched so that that it became

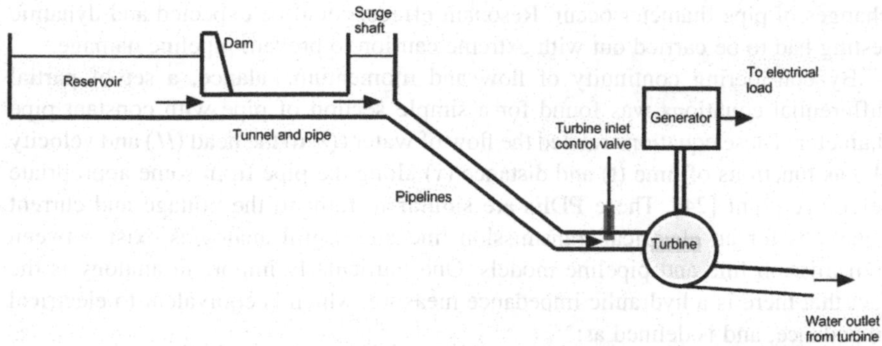

Fig. 7.1 Schematic diagram of the pipeline, turbine and generator system

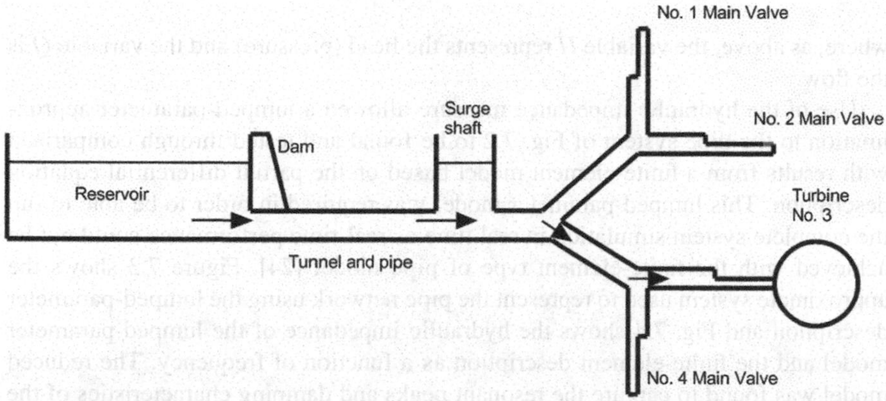

Fig. 7.2 Pipeline schematic diagram showing the geometry of the branching pipeline system. During the tests the main valves for the three turbines were completely closed so that only Turbine No. 3 was in operation and all initial simulation work was carried out for this configuration of the system

the only source of energy for a relatively small, local, electrical distribution network.

Several types of governor were to be investigated in real-time simulation studies, ranging from an existing and relatively slow type of mechanical-hydraulic governor, to several types of proposed fast-acting analogue and digital electronic governors. Although some testing of the real system was allowed for model development purposes, dynamic testing was severely limited due to safety issues associated with the pipeline system. Indeed modelling of the pipeline system in real-time was one of the most important issues that had to be tackled. Disturbances in terms of pressure (head) and flow anywhere within the pipe network are transmitted along the pipes, with possible reflections at all the discontinuities in the system, such as at closed turbine inlet valves, at the surge shaft and reservoir or at the points where changes of pipe diameter occur. Resonant effects could be expected and dynamic testing had to be carried out with extreme caution to prevent pipeline damage.

By considering continuity of flow and momentum balance, a set of partial differential equations was found for a simple section of pipe with constant pipe diameter. These equations related the flow of water (Q) to the head (H) and velocity (V) as functions of time (t) and distance (x) along the pipe from some appropriate reference point [24]. These PDEs are similar in form to the voltage and current equations for an electrical transmission line and useful analogies exist between transmission line and pipeline models. One particularly important analogy is the fact that there is a hydraulic impedance measure, which is equivalent to electrical impedance, and is defined as:

$$Z = \frac{H}{Q} \tag{7.9}$$

where, as above, the variable H represents the head (pressure) and the variable Q is the flow.

Use of the hydraulic impedance measure allowed a lumped-parameter approximation to the pipe system of Fig. 7.2 to be found and tested through comparison with results from a finite-element model based on the partial differential equation description. This lumped-parameter model was required in order to be able to run the complete system simulation in real time as real-time performance could not be achieved with the finite-element type of pipe model [24]. Figure 7.3 shows the approximate system used to represent the pipe network using the lumped-parameter description and Fig. 7.4 shows the hydraulic impedance of the lumped-parameter model and the finite-element description as a function of frequency. The reduced model was found to capture the resonant peaks and damping characteristics of the impedance diagram obtained using the finite-element approach and the accuracy of the lumped-parameter description was judged, by the NHSEB engineers involved in the project, to be acceptable over the frequency range from 0 to 1 Hz.

The lumped-parameter pipeline model was incorporated within the overall system model, which included the hydraulic and mechanical characteristics of the turbine, the electrical and mechanical characteristics of the generator and the electrical load. The main hydraulic servo-motor that controls the turbine guide vanes that function as the turbine control inlet valve was modelled in detail and the computer simulation included an interface to allow electronic control hardware to be incorporated so that new governor designs could be tested using hardware-in-the-loop methods prior to testing on site.

Validation had to depend, in part, on the knowledge and expertise of NSHEB engineers and also experienced operators from the power station in question. Initial face validation of the simulation model involved the real-time simulation of the pipeline, turbine and generator system, together with a model of the existing mechanical-hydraulic governor. The operators, working with NSHEB engineers, highlighted some deficiencies in the plant model. They could detect situations from the "feel" of the real-time simulation which were not typical of the real plant and in

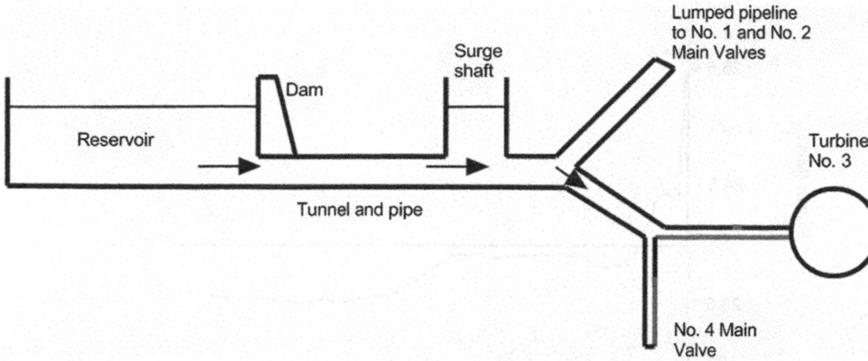

Fig. 7.3 Second approximation used for the development of a lumped parameter approximate model involving a more structurally accurate representation which allows for reflections from the ends of pipes not in use

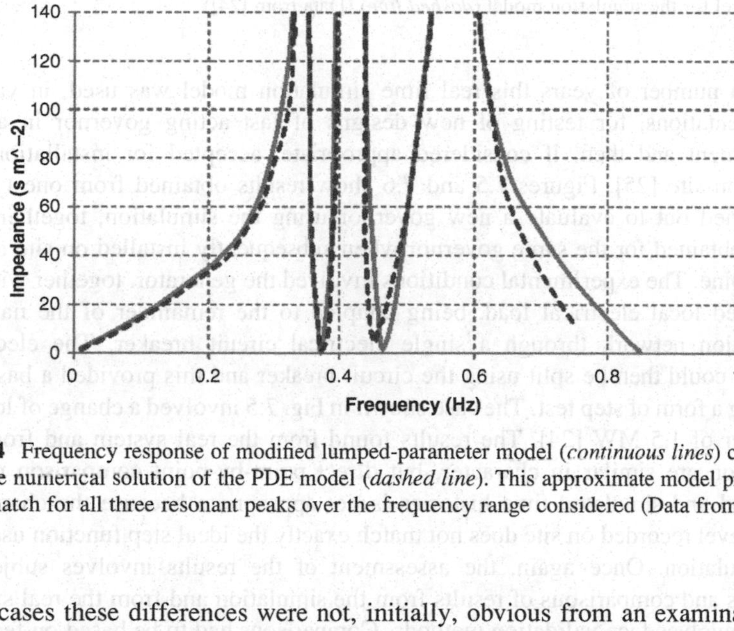

Fig. 7.4 Frequency response of modified lumped-parameter model (*continuous lines*) compared with the numerical solution of the PDE model (*dashed line*). This approximate model provides a good match for all three resonant peaks over the frequency range considered (Data from [24])

some cases these differences were not, initially, obvious from an examination of time history records of plant and simulation variables. Subjective feedback of this kind was useful, for example, in establishing that problems existed within the representations used for the nonlinear mechanical characteristics of the servomotor and guide-vane linkages. Errors in the representation of guide-vane backlash and servomotor saturation and rate limits could lead to small limit cycles occurring for some operating conditions which did not match situations encountered when operating the real system. This face validation procedure allowed changes to be made in the real-time simulation which eventually allowed it to be accepted by the NSHEB engineers and approved for use.

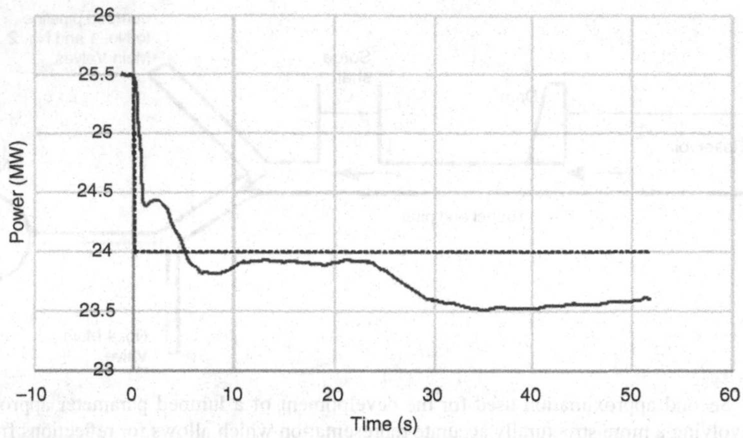

Fig. 7.5 System-splitting test results. The *solid line* shows the time history of generator power output measured on site and the corresponding time history of the ideal 1.5 MW step change in power level for the simulation model (*dashed line*) (Data from [24])

For a number of years this real time simulation model was used, in various implementations, for testing of new designs of fast acting governor in a safe environment and then, if considered appropriate, accepted for installation and testing on site [25]. Figures 7.5 and 7.6 show results obtained from one typical test carried out to evaluate a new governor using the simulation, together with results obtained for the same governor when subsequently installed on site on the real turbine. The experimental conditions involved the generator, together with the associated local electrical load, being coupled to the remainder of the national distribution network through a single electrical circuit breaker. The electrical network could then be split using the circuit breaker and this provided a basis for applying a form of step test. The case shown in Fig. 7.5 involved a change of load of the order of 1.5 MW [24]. The results found from the real system and from the simulation are similar in character, but direct point-by-point comparison of the measured and simulated time histories is not appropriate because the change in power level recorded on site does not match exactly the ideal step function used for the simulation. Once again, the assessment of the results involves subjective elements and comparisons of results from the simulation and from the real system results involves face validation methods. Comparisons had to be based on features such as times to peak value following the imposed change, damping factors and frequencies of oscillations rather than on a detailed comparison of the measured and simulated time histories. What was of particular interest is that the results in Fig. 7.6 show that pressure fluctuations within the pipeline system are acceptable. The transient change of frequency following the applied step change of power was also judged to be acceptable for the governor under test.

It should be noted that the system splitting tests had to be carried out for a variety of different power levels, involving both power import and power export situations,

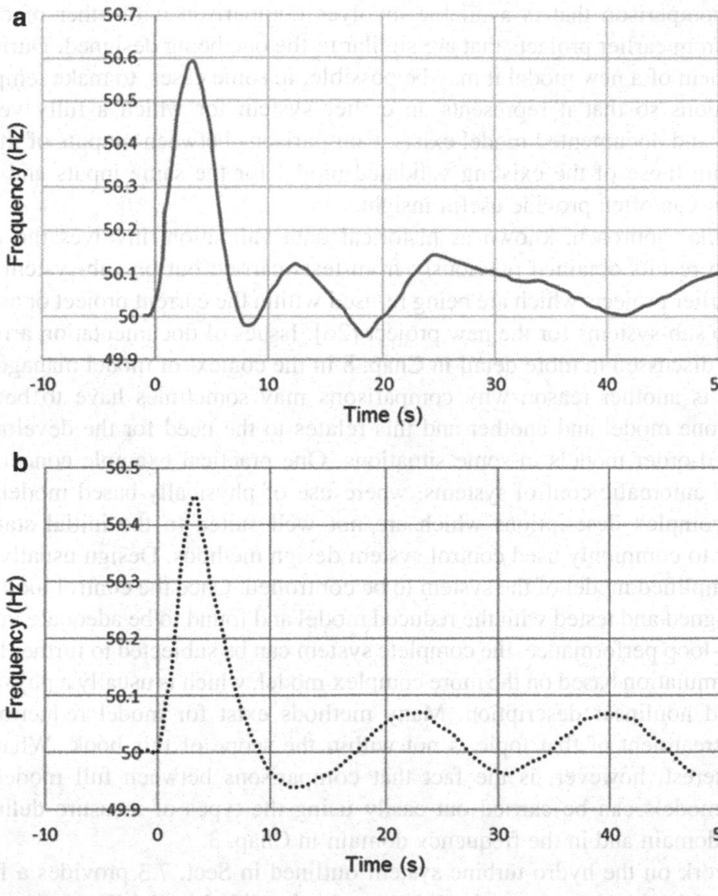

Fig. 7.6 Comparisons of time histories of frequency transient from site test (**a**) and from the simulation model (**b**) for the system-splitting test of Fig. 7.5 with a double-derivative type of electronic governor (Data from [24])

and that repeated trials had to be carried out for each test condition in order to provide all the information needed about the performance of the governor under investigation. These test results and equivalent findings from the simulation model also provided valuable additional information relating to the overall quality of the real-time simulation model.

7.4 Approaches Based on Comparisons with Other Models

In engineering design applications a "real system" only exists in physical form towards the end of the design and development process when a prototype has been built or the system itself exists in physical form. Until that stage is reached the only type of comparison that is available involves comparison with other models of systems from earlier projects that are similar to the one being designed. During the development of a new model it may be possible, in some cases, to make temporary modifications so that it represents an earlier system for which a fully verified, validated and documented model exists. Comparisons between outputs of the new model with those of the existing validated model for the same inputs and initial conditions can often provide useful insight.

A similar approach, known as historical data validation, involves the use of validation results obtained previously from tests carried out on sub-systems used within earlier projects which are being re-used within the current project or are very similar to sub-systems for the new project [26]. Issues of documentation arise and these are discussed in more detail in Chap. 8 in the context of model management.

There is another reason why comparisons may sometimes have to be made between one model and another and this relates to the need for the development of reduced-order models in some situations. One practical example concerns the design of automatic control systems, where use of physically-based models may involve complex descriptions which are not well suited to the initial stages of design or to commonly used control system design methods. Design usually starts with a simplified model of the system to be controlled. Once the control loops have been designed and tested with the reduced model and found to be adequate in terms of closed-loop performance, the complete system can be subjected to further testing using a simulation based on the more complex model, which is usually a physically-based and nonlinear description. Many methods exist for model reduction and detailed treatment of that topic is not within the scope of this book. What is of direct interest, however, is the fact that comparisons between full models and reduced models can be carried out easily using the types of measure defined in the time domain and in the frequency domain in Chap. 3.

The work on the hydro-turbine system outlined in Sect. 7.3 provides a further example of the development and validation of a simplified model. The full model of the pipeline system could not be run in real time to allow testing of fast-acting governors and a reduced model had to be developed which would be considered of adequate accuracy to allow testing of novel forms of speed control system prior to being tested on site.

7.5 Data Sets for Model Testing

Experiments, in both science and engineering, are usually intended to enhance the investigator's understanding of the behaviour of the system being considered. In science, this usually involves hypothesis testing and is most often simply part of the scientific method. In engineering, on the other hand, experimental testing may help in developing mathematical models of existing systems, in estimating parameters of the system about which there are uncertainties, or in providing information that is useful in assessing performance limitations. The data sets gathered from such experiments and tests, although useful for the immediate purposes of the experiment, are seldom appropriate for the purposes of model validation because information is missing about aspects of the experiment, or because the inputs applied do not excite the system in the way that is required for validation. In particular, it is often found that published results from traditional experiments are inappropriate for model validation purposes because of the form of input stimulus used, or because of a failure to record initial conditions of all of the state variables or because of inadequate documentation. For model validation purposes it is best to design experiments that are tailored precisely to the task in hand.

A well-designed validation experiment involves specification of all initial conditions and boundary conditions, applied loads, the form of inputs and measuring equipment used so that few (if any) assumptions need to be made by the modeller [27]. The data gathered from the validation experiment form the standard against which competing models may be compared and it is, therefore, essential to know the limits of accuracy of the experimental data. Only then can predictions from the model be properly assessed.

Errors in experimentation may be classed either as being random errors or systematic errors. The random type of error arises because of issues of sensor precision and measurement noise and such errors contribute to the scatter of data in repeated experiments. On the other hand, systematic errors lead to bias in results which cannot be reduced with additional testing [27]. These errors could arise from sensor calibration inaccuracies, offsets and data acquisition problems.

It is important that those involved with the development of a simulation model should have a role in the design of experiments for model validation and should work closely with those dealing with experiments [27]. This is discussed further in Chap. 8 in the context of model management.

The choice of data sets to be used for testing models raises some interesting issues. Test data used for external validation must clearly be suitable for the intended application of the model. Otherwise it will not be possible to make decisions about the suitability and quality of the model for that application and use of the model could then be unnecessarily restricted. Data sets used for model testing need to be chosen using methods broadly similar to the choice of data sets used for identification, in terms of their spectral properties and amplitude distributions. On the other hand, if any experimental data sets have been used for system identification and parameter estimation purposes during the development of a specific model, it is very important that those data sets should not be used again for validation purposes. However, if any part of a model has been developed using experimental modelling methods it is inevitable that the requirements in terms of test signal amplitude and frequency range for model testing will be broadly similar to the requirements for identification and parameter estimation purposes. Issues that arise depend on whether the model being evaluated is linear or nonlinear in form.

In the case of linear models, it is important to use experimental test records that differ significantly from the records used in the parameter estimation process but are similar in terms of their amplitude and frequency ranges. The choice of experimental records for validation of identified models is discussed in Chap. 10 in the case study relating to helicopter flight mechanics modelling.

Validation experiments should be designed to ensure that all the relevant initial conditions and input data are captured. Simulation data should always be processed in the same way as experimental data so that, for example, any filtering applied to measured data is also applied to the simulated response data.

In the case of a nonlinear model the choice of test conditions is more complex since the system must be excited in such a way that the full range of the significant nonlinearities is covered. The system must also be excited over the entire frequency range of interest. Ideally, what is needed is a way of calculating confidence intervals for model predictions and using these in the model validation process. This is far from straightforward in the case of general nonlinear physics-based parametric simulation models but it should be noted that in some types of non-parametric descriptions, such as Gaussian Process models, additional information of this kind is available (see, e.g., [28]). Also, in the case of linear models and frequency domain methods of experimental modelling, coherence estimates can be used to establish the range of frequencies over which a linear description can be used (see, e.g., [29, 30]). More research is needed to establish better ways for assessing the accuracy of predictions from nonlinear physics-based models and more experience needs to be gained in using techniques such as coherence analysis and Gaussian Process models in applications.

7.6 Validation of Sub-models and Generic Models

Successful use of a model within a library of sub-models depends upon the user having confidence in the verification and validation processes considered when that model was accepted for inclusion in that library. Not only must library models be fully documented so that any questions in terms of the suitability, or otherwise, of a given model for a specific application can be answered easily, but details of the acceptance criteria must also be available.

7.6.1 Library Sub-models

As already pointed out in earlier chapters, the modelling of a real system is an iterative process in which testing and evaluation are of central importance. Consideration of model quality and model validation cannot be separated from the other processes of model development. Validation processes for a sub-model included in a library must be based on the procedures and processes for validation of any other type of model.

7.6.2 *Generic Models*

Some of the general issues that may arise in the validation of a generic model are discussed in [31] and illustrated through problems arising in a specific generic model developed for electro-optic sensor systems projects [32]. Although that generic model was for a specialised application, the methodology is applicable to other generic models.

What was done for that sensor systems model involved the use of existing models for three specific applications and then these special-purpose models were used in testing the generic model. The cases considered involved a thermal imager system model, an infra-red search and track system model and a missile-warner system model. Results from tests on the generic model configured for each of these applications in turn could then be compared with corresponding results from each of the special-purpose models and this allowed confidence in the generic description to grow. As confidence increased new modules could be added to the generic model, but the modified generic description always had to be fully re-evaluated using results from tests carried out on the special purpose models, through a form of regressive testing.

In some situations a model may be needed for a new application and, if a generic model is available, consideration should be given to adapting that generic model structure for this new application. Further testing of the generic model is then necessary and if the validation process fails the user may deduce that either a flaw has been found in terms of the conceptual model for the new application or a limitation has been encountered in the generic model, requiring modifications to allow its capabilities to be extended.

7.7 Special Issues with Distributed Parameter Models

Although much of what has been written in the sections above applies to distributed parameter models as well as to lumped-parameter models there are a number of points that arise with distributed parameter models that require particular attention. Testing the suitability of a distributed–parameter model using experimental data that involves noisy measurements from the corresponding real physical system is inevitably more complex than the verification and validation of lumped-parameter models. One of the key issues relates to the quantities to be compared in the system and the model, the number of sensors to be used for measurements in the real system and the positions of those sensors. Clearly the larger the number of sensors used the better the resolution, but it is immediately obvious that there is a balance to be achieved between resolution and the cost of the experimental programme. It also has to be remembered that in some situations the making of measurements changes in some way the system that is under investigation through, for example, the additional mass of the sensors and the associated wiring or telemetry hardware needed to transmit information back to the investigators. It has been suggested that, in cases where validation is impractical, sensitivity analysis methods should be employed to try to establish which assumptions have the greatest effect on the simulation results [33].

Interest in the problems of validating distributed parameter models has increased significantly in recent years and a large number of publications on this topic have started to appear. There has also been considerable interaction between different fields, with methods adopted in the validation of structural dynamics models being applied in other areas. For example, methods that have been developed by the computational fluid dynamics (CFD) community (see, e.g. [34]) have been discussed in the context of models of ice sheets (see, e.g. [35]).

7.8 Validation of Discrete-Event and Hybrid Models

Although discrete-event simulation models are not the main area of interest in this book they do require some consideration within this chapter. The main focus for validation activities in a discrete-event model relates to the assumptions regarding probability distributions of events. If a certain distribution is defined in the model for a specific discrete variable, the correctness or otherwise of the underlying assumptions need to be tested using data obtained from observations of the equivalent real-world system. For example, a discrete-event simulation of a road junction is likely to involve assumptions about the probability of the time intervals between the vehicles arriving at the junction from each of different directions. These assumptions have to be tested, as do any other assumptions within the discrete-event simulation model.

In the case of hybrid systems involving, for example, an embedded digital processor and associated analogue-to-digital and digital-to-analogue converters, the tasks involved in model validation can often be considered to be broadly similar to those arising in continuous system simulation model validation. However, additional variables may have to be monitored, including the times of occurrence of events within the digital processor and the converters to ensure that the representations of these discrete-time elements of the real system within the hybrid model are adequate.

7.9 Acceptance or Upgrading of Models

Comparisons between the model behaviour and the behaviour of the real system for a specific situation can help to identify potential problems and it is then necessary to perform some analysis of the discrepancies and to propose upgrades for the model. Changes of structure or parameters of the model must be implemented and their significance evaluated systematically using simulation. This may lead to further iterations within the model development cycle.

Parametric changes are usually investigated before any structural changes are considered. Within lumped representations, model parameters are often used to approximate some more complex effect and there are always limits to the range of conditions over which such approximations are valid. The functional validity of a model can sometimes be improved through the tuning of parameters, but this can only be done if the adjustment is within an appropriate range in terms of the parameters. Parameter adjustment using global optimisation methods without regard to known uncertainties and physical limits can give very misleading results.

Often, when parameter values are found that appear physically meaningless, we are dealing with a situation where the model has an inappropriate structure. In this context, establishing the range of frequencies over which discrepancies occur can be very useful and a variety of frequency–domain techniques, such as analysis of coherence and partial coherence can be applied. Deficiencies in model structure are generally more difficult to investigate and rectify but, once dealt with successfully, the upgraded model should have a broader range of applicability.

In some cases, deficiencies in the model can be linked to specific state variables of the model or to particular physical phenomena. If errors between a model variable and the corresponding system variable are found to be correlated with a specific state variable there may well be a problem with the model. A more complex representation of the sub-system associated with that state variable might then be appropriate. Padfield and Du Val [36] have used this type of approach in helicopter flight mechanics model validation. As an example, they have shown that correlation of an output error with helicopter rotor speed suggests that changes are needed within the model for the coupled sub-system involving the engine, drive train and rotor.

Correlation of model errors with derivatives of particular state variables of the model also suggests that a higher-order description may be needed. When regression techniques do not suggest that the errors are linked to specific state variables, or to their derivatives or some linear combination of these, nonlinear combinations of model state variables may have to be considered, but this should be approached, as far as possible, on a physical basis.

7.10 Discussion

External validation, whether of the functional or physical kind, involves two distinct stages. The first of these is concerned with establishing a range of conditions over which a given model can be used for a specific accuracy level. That accuracy level can be defined, generally, in terms of frequency and amplitude. The second stage is concerned with establishing deficiencies in the model and the upgrades that would be necessary in order to achieve a level of performance appropriate for the intended application.

As soon as a working simulation model becomes available, results from it need to be assessed using prior knowledge of the system, information from previous models, comparison of simulation results with results of mathematical analysis carried out on the underlying model and, where possible, results from tests carried out on the real system. If the model testing process is successful and it is judged that the simulation model is appropriate for the intended application, the model may be used until any new evidence becomes available that suggests that that model is in some way deficient.

Figure 7.7 is a block diagram showing the processes of external validation and is broadly similar to Fig. 2.1 of Chap. 2. However Fig. 7.7 but includes more detail in terms of the use of pre-test experimental data, the application of system

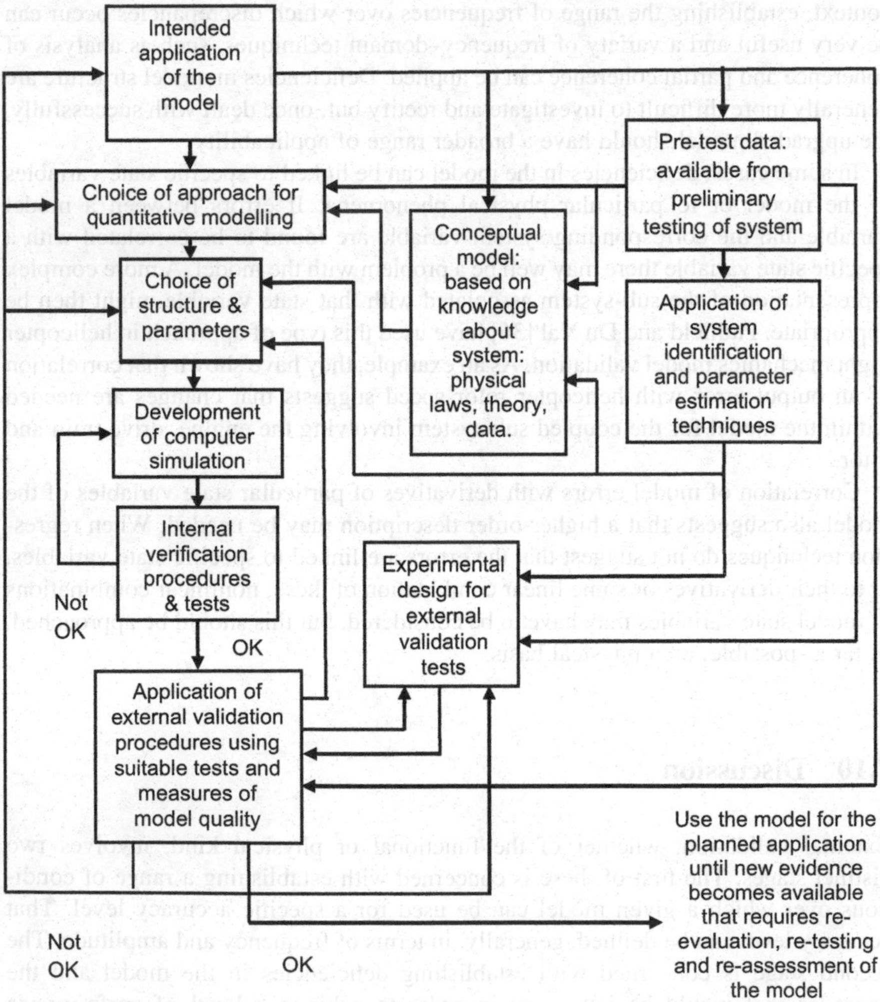

Fig. 7.7 Block diagram showing detailed steps and feedback pathways involved in the process of simulation model development, testing and approval

identification and parameter estimation methods in helping to establish the structure and parameters of the model and the pathways to be followed if it is found at the validation stage that the model quality metrics used to assess the model at the current stage of the validation cycle show that accuracy requirements are not met.

Figure 7.7 also includes the processes of verification but separates these very clearly from the processes of validation. The various feedback pathways within the diagram also emphasise the iterative nature of the whole procedure and it is important to remember that the definitions given for "verification" and "validation" include the words "process of determining" and that both definitions emphasise comparison with references of some kind. For the internal verification process this reference might be an analytical solution for a specific simple case, while for external validation the reference might be a measured or observed set of responses for one or more variables of the real-world system.

For such comparisons an appropriate estimate of the accuracy of any measured quantities must be available. Indeed, clear statements about the level of acceptable error must be established on an a priori basis for both the internal verification and external validation processes. Figure 7.7 further emphasises the importance of the knowledge base relating to the real system, the role of experimental procedures not only for the estimation of parameters but also for the choice of model structure and the importance of experimental design in the external validation process.

References

1. Oberkampf WL (2007) Predictive capabilities in computational science and engineering. Presented at OASCR Applied Mathematics PI Meeting, Lawrence Livermore National Laboratory, 22–24 May 2007. http://science.energy.gov/~/media/ascr/pdf/workshops-conferences/mathtalks/Oberkampf.pdf. Accessed 10 June 2015
2. Anderson J, Papachristodoulou A (2009) On validation and invalidation of biological models. BMC Bioinforma 10:132. doi:10.1186/1471-2105-10-132, http://www.biomedcentral.com/1471-2105/10/132. Accessed 10th June 2015
3. Gore R, Diallo S (2013) The need for usable formal methods in verification and validation. In: Pasupathy R, Kim S-H, Tolk A et al (eds) Proceedings of the 2013 winter simulation conference. IEEE, Washington, DC, pp 1257–1268. doi:10.1109/WSC.2013.6721513
4. Heitmeyer CL (2007) Formal methods for specifying, validating and verifying requirements. J Univ Comput Sci 13(5):607–618
5. Gustavsson I (1972) Comparison of different methods for identification of industrial processes. Automatica 8(2):127–142
6. Prajna S (2003) Barrier certificates for nonlinear model validation. In: Proceedings 42nd IEEE conference on decision and control 9–12 Dec 2003, vol 3. IEEE, Washington, DC. pp 2884–2889, doi:10.1109/CDC.2003.1273063
7. Tischler MB, Remple RK (2012) Aircraft and rotorcraft system identification, 2nd edn. AIAA, Reston, VA
8. Knudsen M (2006) Experimental modelling of dynamic systems: an educational approach. IEEE Trans Educ 49(1):29–38
9. Gong M, Murray-Smith DJ (1993) Model reduction by an extended complex curve-fitting approach. Trans Inst Meas Control 15(4):188–198

10. Balci O, Sargent R (1984) Validation of simulation models via simultaneous confidence intervals. Am J Math Manag Sci 4:375–406
11. McFarland J, Mahadevan S (2008) Multivariate significance tests and model calibration under uncertainty. Comput Methods Appl Mech Eng 197(29–32):2407–2479
12. Huynh DPB, Knezevic DJ, Patera AT (2012) Certified reduced basis model characterization: a frequentistic uncertainty framework. Comput Methods Appl Mech Eng 201:13–24
13. Rosenberg JR, Murray-Smith DJ, Rigas A (1982) An introduction to the application of system identification techniques to elements of the neuromuscular system. Trans Inst Meas Control 4 (4):187–201
14. Butterfield MH, Thomas PJ (1986) Methods of quantitative validation for dynamic system models-part 1: theory. Trans Inst Meas Control 8:182–200
15. Cameron RG (1998) Model validation by the distortion method: linear state space systems control theory and applications. IEE Proc D 139(3):296–300
16. Thomas PJ (1999) Simulation of industrial processes for control engineers. Butterworth-Heinemann, Oxford
17. Cameron RG, Marcos RL, De Prada C (1998) Model validation of discrete transfer functions using the distortion method. Math Comput Model Dyn Syst 4(1):58–72
18. Gray GJ, Voon LK, Murray-Smith (1997) Application of the distortion method for model validation. In: Troch I, Breitenecker F (eds) Proceedings 2nd MATHMOS VIENNA IMACS symposium on mathematical modelling February 1997. Argesim, Vienna, pp 1033–1038
19. Butterfield MH, Thomas PJ (1987) Quantitative validation of dynamic models for use in fast reactor safety assessments. In: Proceedings international conference on science and technology of fast reactor safety, Guernsey (UK), 12–16 May 1986. British Nuclear Energy Society, London, pp 145–152
20. Butterfield MH, Thomas PJ (1986) Methods of quantitative validation for dynamic system models-part 2: applications. Trans Inst Meas Control 8:201–219
21. Butterfield MH (1990) A method of quantitative validation based on model distortion. Trans Inst Meas Control 12:167–173
22. Kleijnen JPC (1995) Verification and validation of simulation models. Eur J Oper Res 82:145–162
23. Chattee E, Saam NJ, Möhring M (2000) Sensitivity analysis in the social sciences: problems and prospects. In: Suleiman et al (eds) Tools and techniques for social science simulation, chapter 3. Physica Verlag, Heidelberg
24. Bryce GW, Foord TR, Murray-Smith DJ, Agnew P (1976) Hybrid simulation of water turbine governors. In: Crosbie RE, Hay JL (eds) Simulations councils proceedings series 6(1), Simulation Councils, La Jolla, pp 35–44
25. Bryce GW, Agnew PW, Foord TR et al (1977) On-site investigation of electrohydraulic governors for water turbines. Proc IEE 124(2):147–153
26. The Mitre Corporation (2014) Verification and validation of simulation models. In: Mitre systems engineering guide, pp 461–469. www.mitre.org/publications/technical-papers/the-mitre-systems-engineering-guide. Accessed 10 June 2015
27. Hemez FM (2004) The myth of science-based predictive modelling. In: Proceedings foundations'04 workshop for verification, validation and accreditation (VV&A) in the 21st century, Arizona State University, Tempe, Arizona, 13–15 October 2004. Report LA-UR-04-6829, Los Alamos National Laboratory, USA
28. Kocijan J, Girard A, Banko B et al (2005) Dynamic system identification with dynamic processes. Math Comput Model Dyn Syst 11(4):411–424
29. Tischler MB (1996) System identification for aircraft flight control development and validation. In: Tischler MB (ed) Advances in aircraft flight control. Taylor and Francis, London, pp 35–69
30. Tischler MB, Remple RK (2006) Aircraft and rotorcraft system identification. AIAA, Reston
31. Smith MI, Murray-Smith DJ, Hickman D (2007) Verification and validation issues in a generic model of electro-optic sensor systems. Def Model J Simul 4(1):17–17

32. Smith MI, Murray-Smith DJ, Hickman D (2007) Mathematical and computer modeling of electro-optic systems using a generic modeling approach. J Def Model Simul 4(1):3–16
33. Walmsley CW, McCurry MR, Clausen PD et al (2013) Beware the black box: investigating the sensitivity of FEA simulation to modelling factors in comparative biomechanics. PeerJ 1:e204, http://dx.doi.org/10.7717/peerj.204}. Accessed 10 June 2015
34. Rizzi A, Vos J (1998) Towards establishing credibility in computational fluid dynamics simulations. AIAA J 36(5):668–675
35. Thompson, DE (2005) Verification, validation and solution quality in computational physics: CFD methods applied to ice sheets, NASA/TM-2005-213453, NASA Technical Reports Server, 37 pp
36. Padfield GP, Du Val RW (1991) Application areas for rotorcraft system identification: simulation model validation. In: AGARD Lecture Series 178, Rotorcraft System Identification, 12.1-12.30, AGARD, Neuilly-sur-Seine, France

Chapter 8
Management Issues Within Simulation Model Development and Testing

In order to fully exploit the important intellectual property that simulation models represent and to obtain the full benefits that models can provide it is essential that there should be an appropriate strategy for the management of models and for their verification and validation. The aims of modelling and computer simulation activities are different in different application areas but detailed prior knowledge of the real system and understanding of the requirements for a simulation model are always important. Models can be particularly helpful within a multi-disciplinary environment in that they form a significant part of the knowledge base used by all the team members, whatever their backgrounds and wherever their locations may be. A well-managed and fully-documented set of models, computer simulations and associated databases is a vitally important asset, not only for the immediate project for which the models have been developed, but also for future projects.

8.1 The Need for Management Procedures

Within many organisations, at present, simulation models are applied without any proper quality assurance mechanisms being put in place. Those models may be developed within that organisation, or recovered from earlier projects, or obtained from other organisations or purchased from consultants. In all cases strategies need to be adopted to ensure that model quality issues are properly addressed and that tools used for modelling and simulation activities are appropriately managed.

This applies in scientific and medical applications just as much as in engineering. Modelling and simulation methods need to be applied with care to avoid undermining the creative output and efforts of complete project teams. Current practice in simulation modelling often fails to recognise the central importance of testing, verification, validation and documentation. This is different from accepted procedures in most other software-dependent fields, where rigorous testing,

D.J. Murray-Smith, *Testing and Validation of Computer Simulation Models*,
Simulation Foundations, Methods and Applications,
DOI 10.1007/978-3-319-15099-4_8

documentation and version control policies are usually applied because they are recognised as being cost effective and helpful in the development process.

Recent years have seen much more emphasis on the use of libraries of sub-models and also, in some fields, on the use of generic models that can be applied to a range of different, but related, projects. Thus, confidence in a prediction made using a large and complex model often depends, to a significant extent, on confidence that can be placed in sub-system models. This is particularly important where sub-system models can be tested experimentally since comprehensive testing of sub-models allows confidence to be established at that level first and then extended gradually to less well-defined situations involving the complete system model over a range of conditions. It is becoming clear that more effort needs to be directed towards finding better ways of developing and maintaining libraries of validated simulation models and commonly used sub-models. This is important if we are to fully exploit the benefits of model re-use and the development of reliable generic models.

In design situations the main aim in using modelling and simulation is to ensure that there will be as few unwelcome surprises as possible when the system under development is finally built, tested and put into service. The complete elimination of problems is unlikely, but detailed testing of models used as the basis for design greatly increases the chances that the resulting system performance will be found to be acceptable at an early stage in the design and development cycle. Clearly, budget constraints and development timescales are likely to make it impossible to validate a model for every component and sub-system defined in the validation hierarchy outlined in Chaps. 2 and 7. However, applying sensitivity analysis and expert judgement in a constructive fashion, even at an early stage when the model is far from fully developed, can provide valuable insight that can help reduce unnecessary activities and focus attention on elements of the model where uncertainties are most important, thus reducing the total expenditure. Estimation of uncertainties at the component and sub-system levels in this way allows their effects to be estimated at the top level of the validation hierarchy.

In the development of a new engineering system, experimental data from the complete system is unavailable at the design stage. In some cases historical data from earlier systems of a similar kind can be helpful in evaluating the model of the new system. Successful application of this approach depends on maintaining and updating previous models as the software environment changes, providing good documentation for models of the earlier systems and also continuing to make available the experimental data and tests used to evaluate those previous models.

A key aspect of model management for a simulation application involves establishing a model verification, validation and accreditation plan. This should be established before other aspects of the project are started. The plan must establish quantitative requirements for model validation which indicate clearly the extent of validation required for each element within a complex model, including sub-models that require validation. The plan should guide the user through each phase of the verification, validation and accreditation (VV&A) process and is different and quite distinct from the model specification. The plan should also describe the methods to be used and indicate relevant resources available from previous projects (such as established and validated models of any sub-systems considered previously).

In applications where predictive validation is to be used it is also important to determine how much data may be needed in order to meet the requirements for validation. A trade-off then becomes possible between the quantity of validation data needed and the validation requirements since the cost of obtaining such data usually is of key importance in terms of the overall cost of a modelling and simulation project. Another important fact that is often neglected is that it is not always essential to validate every sub-model since, in some cases, adequate information may be available from previous projects or from the manufacturer of the components involved in those sub-systems.

The plans for model development and for verification, validation and accreditation should also make proper allowance for the fact that the model and the VV&A procedures must often be developed in parallel. This is due to the fact that, as already mentioned, the development process for a model is inherently iterative. Validation data has to be gathered and analysed and decisions can be made during the model development cycle to make changes in the model as a result of this analysis process. The VV&A plan should always identify sub-models that may be validated independently of other elements but, in some cases, there may be problems in the design of tests to provide test data that allow a sub-model to be validated over the complete operating range for that element. In such cases the initial validation may be performed over part of the full operating range, with further investigation deferred until the sub-model is operating in conjunction with the other model elements.

Although the management plan may initially suggest some specific methods of model validation, flexibility is needed in terms of approaches used and a variety of methods may need to be applied at different stages in a project. For example some preliminary face validation may be used initially and, following a successful outcome from this initial expert review a more quantitative approach involving predictive validation methods may be used.

Those involved in development of a model should be able to contribute fully to the design of validation tests. Simulation specialists should collaborate closely with experimentalists responsible for the measurement work and test rigs. Prior to a model being available that has gone through a full process of validation, any preliminary simulation results must be treated with extreme caution, but useful information from early versions of a simulation model may help in experimental design. For example, sensitivity studies can help with test input design and initial information from the modelling and simulation work can be helpful, in a general way, for those involved in planning of tests and other experimental work. Similarly, those involved in the conduct of experiments should be able to provide valuable information about experimental errors. Along with information about model uncertainties, this can be very helpful in the model evaluation process.

Although management should ensure that there is close cooperation between those responsible for model development and those involved in testing and experimentation on the real system, it is important that those involved in model development do not have direct access to the detailed experimental results until they have used the simulation model to make predictions for the conditions considered experimentally. Ideally, the results found from simulation and from experimental investigations should be evaluated through some sort of blind comparison. This might involve the experimentalists providing the simulation team with measured input data from the experimental programme but the experimental group would retain the measured system response records initially. The simulation team would then use the input data provided by the experimentalists to make predictions of outputs and these could be compared, by both teams working together, with the corresponding experimental findings. This allows fair comparisons to be made. Any further tuning of the model as a result of the failure of model predictions to match experimental outputs should be part of a further full iteration in the modelling development process. Ad hoc adjustment of the model to match an experimental finding, without subsequent evaluation and testing, does not constitute validation and should be actively discouraged through appropriate management and planning.

In the context of engineering design and development applications, a key to good management of the simulation model development process lies in establishment of a properly structured working group which meets regularly to review progress and make recommendations about issues such as the collection of additional validation data. Membership of the group should include at least one from the system development team, one involved in the development of the model and one who has experience of conducting tests. In the case of government contracts a member of the working group should represent the sponsoring government organisation. In other contexts this could be a representative of the customer organisation or of the user community (e.g. an experienced operator of the type of equipment being developed).

Project management for simulation model VV&A must also involve careful consideration of the capabilities and resources required for the verification and validation activities. The availability of system experts, simulation experts and people with system test/data analysis expertise is obviously important, as is the availability of appropriate computing resources and software [1, 2].

Another issue that affects simulation tools as much as any other software is a tendency for software quality to deteriorate over time if strong management procedures are not used. Transfer of personnel who previously had specific responsibilities for development and maintenance of particular models is one obvious source of difficulty as vital knowledge and experience may then be lost. Changes to programs that are not fully documented is another obvious example. If there is inadequate documentation, and those who developed or changed programs are not available for consultation about the software, it is important to have some systematic way of re-establishing the models and associated tools in a usable form. This is recognised as a form of "reverse engineering" and methods for overcoming difficulties of this kind have been attracting attention in recent years. One relevant publication is by Birta et al. [3] and this presents concepts of reverse engineering tools which can overcome problems associated with a lack of documentation of existing simulation programs and thus provides assistance in gaining a proper understanding of the organisation and usage of the software. A toolkit is described for implementation of these concepts of reverse engineering for SLAM II simulation programs.

8.2 Tools for the Management of Simulation Models

Tools for model management should allow users to focus attention on model testing and quality issues. Possible integration of simulation tools with other tools for analysis and design is also important. Good software for model documentation and for the overall management of models is especially important when the simulation models are likely to be re-used.

User-friendly and reliable techniques for version handling are essential for large models, especially if they are being developed by a team rather than by one individual. As suggested by Brade [4], this and other aspects of the management system should form elements of the iterative process for model development in which verification, validation and documentation are vital components.

As pointed out in Sect. 1.6, interactions between simulation models of different types can be an important issue for some types of application, such as in the design and manufacture of complete aircraft or aircraft sub-systems. In such applications data transfer between the different simulations must be free from errors and it is also important to ensure that, whenever design changes are made, all the relevant models are updated at the same time. Data transfer and model updating form part of the internal verification process but require active management procedures if they are to be implemented successfully.

8.3 Simulation Model Documentation and the Use of Model Libraries

A model of proven fitness, together with good documentation, often provides a starting point for a new application, whereas poorly-documented models with unknown validity are unlikely to be of assistance. This can also be important when a model of proven quality is to be used in a different way. Although such models are likely to require modification for a new application an existing validated model often provides a good starting point. This requires information about the model requirements for the earlier model, its assumptions and constraints, the range of validity and details of the previous application.

Awareness of model limitations inevitably fades with time and well-structured and accurate documentation is essential for the developer, even if others are not immediately involved. Attitudes towards documentation vary greatly from one person to another and between different organisations. It is clear that, in some cases, projects are completed with little or no documentation in place while, in others, much time and effort is put into creating a well-ordered on-line documentation system. The difference often depends on the approach within the organisation in question to quality assurance procedures in general.

Experience suggests that models developed on a one-off basis, are not always subjected to rigorous validation and documentation procedures. In contrast, well-maintained libraries of re-usable models can often be of great assistance and can help to avoid unnecessary model development work in new applications. In general, whether or not a model is likely to be re-used, information recorded about a model or sub-model should always include:

- the purpose of the model, its application and the associated requirements in terms of overall accuracy,
- sources of funding for the development of the model and the role of the organisations concerned,
- the assumptions and simplifications inherent in the model and any limitations that result from these,
- equations of the model and their sources, variables (including inputs and outputs), parameters etc.,
- details of tests performed on the real system for the purposes of model development, including results of any system identification and parameter estimation techniques applied and including data files used,
- computer simulation code for the model (or simulation block diagram in the case of bock diagram based simulation tools), together with any other model initialisation and set-up information,
- details of verification procedures used to check the consistency of the computer-based simulation model with the mathematical and conceptual models,
- details of validation procedures and tests for sub-models as well as for the complete simulation model,
- model variations developed in the course of the project, together with reasons for accepting or rejecting each model, together with statements about any uncertainties and the range of applicability for accepted models.

It is important to note that the model development process is not complete when a model or sub-model is accepted for the planned application (or "accredited" in more formal situations). Neither can the development be considered as finished when a model is accepted for inclusion in a model library. The process continues as long as information is being gathered from the corresponding real system and goes on, ideally, throughout the duration of the project. In engineering applications this may mean that modelling activities continue in some form throughout the whole life cycle of the real engineering system or product and must take account of modifications applied to the system or product once it is in service. In scientific applications models may continue to be developed as knowledge grows about the corresponding real system. Documentation procedures must therefore allow for ongoing processes of model refinement and should always make allowance for possible updating or modification of the model and of the associated software. "Regressive" testing of simulation models within the iterative procedures of model development is vitally important, just as it is in the development and maintenance of all other kinds of software (see, e.g., [5]). Consequently, both for scientific and engineering applications, model documentation may need to be maintained, simultaneously, for a number of different versions of a given model.

Understanding of a given model and of its limitations should increase steadily with time. For most applications the modelling process can never be considered completed, especially if there is a possibility of re-use of a model, or parts of it, in a future project. Maintaining the quality of documentation requires considerable self-discipline in the case of an individual modeller and appropriate encouragement and helpful management structures are important within larger organisations where modelling and simulation activities are based on teams.

Documentation should not only provide a comprehensive description of the model in question and the related software, but should also allow for the needs of first-time users of a model. For example, diagrams consistent with documentation relating to the corresponding real system should be provided. Methodologies adopted should also allow features within an executable simulation model to be related back to the original set of model requirements [4].

Some simulation tools that were popular and widely used two or three decades ago would now be regarded as being very inflexible and would be seen now as providing rather poor facilities for model entry, for running and for interactive control of the simulation. Also, such tools provide limited facilities for documentation with the result that model documentation often consists only of comment lines within the code or annotations within graphical block-diagram based models. Many simulation and modelling tools developed in recent years place far more emphasis on good model management and on efficient user-interaction. The overall computing environment within which the simulations are built and operated may also allow access to more effective documentation systems. Many modern simulation tools provide access to available libraries of sub-models and permit the creation of new libraries and it is now recognised that methods involving object-oriented software environments offer advantages for the development of re-usable and extendable models.

Brade [4] has suggested that, to generate effective documentation at an acceptable cost, software should allow changes to be recorded automatically throughout the model development process. When compiled manually, such a record is often ineffective, usually incomplete and often results in models that contain significant errors.

It is also important to point out that unless the code and algorithmic checking processes have been performed properly in the verification stage, with documented evidence, it is inappropriate to move forward to validation. As has been stressed in earlier chapters of this book, verification and validation methods can never prove that a simulation model is suitable for an application but simply provide evidence on the basis of which decisions can be made about the possible use of the model. If such decisions are favourable the model may then be applied, probably with some limitations, until further evidence emerges that indicates the presence of additional and previously undetected problems.

As outlined in Chap. 7, a combination of graphical methods, deterministic measures and statistical measures for the evaluation of simulation model output can be helpful in identifying problems and, in future, more highly automated methods of error detection may be available to support this. Brade and Waldner have discussed a possible approach to detect automatically any violations of desired model behaviour [6]. Although such developments could lead, eventually, to automated approaches to external validation and are undoubtedly of interest, it is still generally accepted that automated methods cannot replace the currently used subjective approaches involving human intervention based on face validation techniques.

The US Department of Defense [7] has defined the minimum set of items to be documented as part of the VV&A process. These include obvious information such as the name of the person or organisation responsible for the VV&A activities, information about the dates when the work was carried out and the version or release number of the simulation. Other information to be included in the documentation is broadly consistent with the items specified in the list provided above, with the addition of a statement of the verification and validation outcomes.

In the case of accreditation, there must be a summary of the results of the accreditation assessment and a record of the accreditation decision. There are five possible accreditation recommendations [8]. These are:

- The model or simulation may be used as described
- The model or simulation may be used as described, with limitations
- The model or simulation may be used as described, with modifications
- The model or simulation should not be used for this application and requires the application of further procedures in terms of verification and validation before it can be re-considered for accreditation
- The model or simulation should not be used for this application.

Most models satisfy the second of these possible outcomes and the limitations have to be properly defined and documented. Users must be made aware of the limits within which a model may be used and this is important both with complete system models and with sub-models.

Like other models, sub-models included within a library and generic models must be fully supported by documentation. Without good documentation, which must include details of the purpose and limitations of each model, such libraries are of little value. The information needed about a library sub-model or a generic model is basically the same as the information required to document any other model. It should include the theory used in the development of the model (with sources referenced properly), the full set of assumptions made, the testing and validation procedures used, information about the range of applicability and the reasons for accepting it as a sub-model or generic model within the library. In some commercial or defence-related application areas libraries may involve precompiled sub-models for which full source code is not made available to users. Particularly comprehensive documentation is necessary in such cases as users have no way of investigating the internal organisation of the sub-models directly. Ideally what is needed is an archiving system that can capture and store all aspects of a design, including models with all the assumptions, variables and simulation code, in an appropriate and scalable electronic format [9].

As well as re-use and adaptation of existing models within new projects there is another way in which a model developed for design purposes can be re-used. This occurs in cases where a validated model, developed and used for design, is used again within a simulator for training personnel to operate the real system or for exploring how to operate the system more efficiently or more effectively. Through this the system performance may be improved or maintenance costs reduced. Validated models can also be used to help in decision-making, in working out how best to deal with faults and in avoiding potentially dangerous emergency situations. In some industries, such as the paper industry, dynamic simulation techniques are used mainly for operator training and for system optimisation [10]. Assessment of the costs incurred in model development must take full account of the benefits that come from these applications, which may appear to be of secondary importance but can be very valuable.

One interesting suggestion relating to model documentation is that it should always be divided into two sections. One of these would involve non-technical documentation and would be accessible to all with a direct or indirect interest in the model. That section would provide an overview of the model, its purpose, intended application, structure, equations and parameter values, along with details of the verification and validation procedures, but would not necessarily provide all the information needed to reproduce the model and the associated simulation software. The other section of the documentation would come under the heading of technical documentation and would involve the full details of the model and would include the complete simulation and all the information needed to run it. From the technical documentation an informed reader with modelling and simulation experience should be able to reproduce all published results. The benefit of splitting the documentation in this way, into what might be termed "public" and "confidential" sections, is that people outside the organisation that developed the model could be given general information about it, together with an outline of the intended purpose and some information about model limitations but would not necessarily be given full access to every detail [11].

8.4 Benefits Versus Costs of Model Management Procedures

One important aspect of model validation, too often neglected by academics, concerns the costs of introducing systems for model management and documentation and, especially, the costs of applying model validation techniques. It should be clear, from the previous discussion in this chapter and in earlier chapters of this book, that the benefits obtainable from applying verification and validation procedures and of introducing appropriate constraints in terms of version control and documentation can be considerable. On the other hand, such policies do have to be properly costed and a balance has to be achieved, in each case, in terms of those benefits versus the costs.

The key issue is to ensure that the model is fit for the intended purpose. The most obvious way in which a model might be judged to be inappropriate arises when the model behaves in ways that do not match the behaviour of the real system. However, a model could also be inappropriate if it describes the real system in ways that are irrelevant for the application. This could, for example, apply to a model that has been defined using requirements that involve a frequency range that extends well above or well below the frequencies of interest for the intended application. Testing of a model therefore always has to take account of the requirements for which it is being developed. Carrying out validation over a range of conditions which is wider than those needed for the application could be very wasteful in terms of time and human effort and could add significantly to the cost.

The best way of controlling the costs of validation is through linking the verification, validation and accreditation plan into the more general requirements analysis document which details the model purpose, provides a top-level model design and defines the strategy for development of the model. An approach of this kind allows an overall project plan to be created which includes the estimated effort and time-scale at each stage and may indicate how the tasks involved in the model development process can be split between different individuals or even teams (if appropriate). Although it may appear, initially, that the introduction of model management systems with project plans, version control and documentation standards may extend the duration of the model development process and have substantial additional manpower costs, this should actually lead to the production of a model that is fit for purpose in a shorter period of time.

Since model development processes are almost always iterative, the level of model detail and the testing procedures for model validation are likely to be refined in a stepwise fashion using the same broad set of requirements throughout. This allows confidence to be built up about the fitness of the model for its intended application while the overall cost of the modelling procedure, including validation, is kept under close scrutiny. A low confidence level in terms of a system model that is being used for design leads, inevitably, to a lack of confidence in terms of the performance of the first prototypes. Prototypes that do not meet performance specifications inevitably lead to costly and time-consuming changes of hardware and software within the system being developed. Although the performance of a prototype is closely linked to the quality of the models used in its design, there is no point in developing a model that is better than is necessary for the intended application as that also leads to additional costs.

One possible reason for attempting to extend the range of conditions for which validation is carried out beyond those that apply to the project immediately in hand relates to re-use and the possible benefits of a more generic model. If models are to be re-used the costs of model development may be viewed as being divided between several projects. This is an issue which must clearly be a matter for discussion in each case when a new model is being developed and the problems and limitations associated with the validation of generic models must be fully understood [12].

It is difficult to obtain firm information about current practices, especially within companies and other commercial organisations, but there are some examples of relevant studies. One interesting and thought-provoking paper on system modelling and simulation as applied in the chemical industry has been published by Foss et al. [13]. The modelling process is discussed in detail in that paper, using information gathered from 16 experienced modellers working in the chemical industry in organisations based in two different countries. The modellers had been interviewed following distribution of information explaining the overall objectives, providing a set of questions to be considered during the interview and describing how interviews would be conducted. Interviewers followed an agreed and standardised procedure. The topics of verification, validation and documentation received a significant amount of attention. The paper includes a number of suggestions for improving modelling technology and includes recommendations for developments in terms of advanced modelling tools.

Although the paper by Foss et al. [13] is based on the chemical industry and the conclusions represent the views of a relatively small number of modellers, it must be recognised that these are people with substantial industrial experience gained over a period of at least 10 years. This study highlights the possible benefits from following good practice in terms of model requirements definition and the adoption of a systematic approach to model development and testing. Although the study applies directly to one industrial sector it could have considerable relevance in other fields and the findings are reflected in information presented in other chapters of this book. One may question the validity of such investigations in that those taking the time to respond may be biased in favour of systematic procedures for model development, especially for verification, validation, model re-use and documentation. However, examination of the available evidence from this particular study undoubtedly provides valuable insight regarding views from industry.

A second study, which focussed more on the aircraft industry, relates to the use of system identification methods in helicopter flight testing [14]. Although that investigation dates from about 1990, the main findings are believed still to be valid. It involved analysis of a questionnaire to industry soliciting views on the use of system identification and parameter estimation techniques for applications such as model validation. A total of eight companies in the USA and Europe responded and the overall results show considerable interest in the use of the system identification approach, although there was also some cautious scepticism about results and the need for a physically-based interpretation at all times. A mix of quantitative analysis of system identification and parameter estimation results with subjective interpretation and face validation techniques appears to be the approach most widely used in model testing. Issues of model structure were given particular emphasis in the responses. The study provides an interesting glimpse of attitudes and expectations within one specialised part of industry. Judging from available published information, the situation seems unlikely to have changed significantly in the period since the study was completed. It also seems likely that that this study from the helicopter industry reflects the views of people in other areas of engineering to the use of system identification techniques for model development and validation purposes.

Cost prediction for the use of a modelling approach to engineering design is inevitably difficult but this has been considered for some large projects in the aerospace and defence sectors and receives attention in a useful document by Pace [15]. However, he points out that information about the costs of modeling and simulation activities is seldom shared and he suggests that much more information about cost and resource requirements needs to be gathered and made more widely available to facilitate development of reliable costing procedures. Another important issue that is often forgotten is that, in practice, it may be difficult to maintain a model over the complete life-cycle of the system or product because advances in technology may make the original simulation code, written in the context of a particular computing environment, obsolete. Good documentation can help greatly in dealing with problems of this kind but models can only be maintained properly if they are seen to be important, either in economic terms or in terms of their future potential for further research. It appears that some engineering organisations and companies are now investing in technology groups which, among other remits, are being given responsibility for maintaining models that are likely to be important in the future and may be re-used or developed further, along with the relevant documentation systems [13]. This may lead, sometimes, to the development of in-house model libraries or generic models. This type of development, inevitably, has its own costs but can lead to savings in the long term, if carried out in an appropriate fashion with adequate documentation.

References

1. The Mitre Corporation (2014) Verification and validation of simulation models. In: Mitre systems engineering guide, pp 461–469. www.mitre.org/publications/technical-papers/the-mitre-systems-engineering-guide. Accessed 24 Feb 2015
2. Balci O, Glasgow PA, Muessiny P et al (1996) Department of defense verification, validation and accreditation (V. V. & A.) recommended practices guide, defense modeling and simulation office, (1996). Defense Modeling and Simulation Office, Alexandria, VA
3. Birta LG, Abou-Rounia AO, Ören TI (1992) Reverse engineering in the simulation lifecycle. SAMS 9:69–84
4. Brade D (2003) A generalized process for the verification and validation of models and simulation results, dissertation submitted for award of the degree Dr rer. nat., Fakultät für Informatik, Universität der Bundeswehr München, Germany
5. Kit E (1995) Software testing in the real world. Addison Wesley, Harlow

6. Brade D, Waldner C (2003) Automatic detection of behavioural specification violations. In: Proceedings of the 03 European interoperability workshop, Stockholm, Sweden. Simulation Interoperability Standardization Organization
7. US Department of Defense (2009) DoD modeling and simulation (M&S) verification, validation, and accreditation (VV&A). U. S. Department of Defense Instruction Number 5000.61. December 9, 2009. http://www.dtic.mil/whs/directives/corres/pdf/500061p.pdf. Accessed 1 Mar 2015
8. Cook DA, Skinner JM (2005). How to perform credible verification, validation and accreditation for modelling and simulation. J Def Softw Eng 18(5):20–24
9. Famme JB, Gallagher C, Raitch T (2009) Performance based design for fleet affordability. Nav Eng J 12(4):117–132. doi:10.1111/j.1559-3581.2009.00233.x
10. Kammel G, Voigt HM, Neβ K (2005) Development of a tool to improve the forecast accuracy of dynamic simulation models for the paper process. In: Kappen J, Manninen J, Ritala R (eds) Proceedings of model validation workshop, Oct. 6, 2005, Espoo, Finland. VTT Technical Research Centre, Finland
11. Eddy DM, Hollingworth W, Caro JJ et al (2012) Model transparency and validation. A report of the ISPOR_Smdm Modeling Good Research Practices Task Force-7. Med Decis Making 35 (5):733–743
12. Smith MI, Murray-Smith DJ, Hickman D (2007) Verification and validation issues in a generic model of electro-optic sensor systems. Def Model J Simul 4(1):17–17
13. Foss BA, Lohmann B, Marquhardt W (1998) A field study of the industrial modelling process. Model Identif Control 19(3):153–174
14. Padfield GD (1991) The role of industry. In: Rotorcraft system identification, AGARD advisory report 280 (AGARD-AR-280), Section 3.3, AGARD, Neuilly-sur-Seine, France
15. Pace DK (2004) Modeling and simulation verification and validation challenges. John Hopkins APL Tech Dig 25(2):163–172

Chapter 9
Case Study: Development and Testing of a Simulation Model of Two Interconnected Vessels

9.1 Introduction to the Case Study

Systems involving vessels containing liquid arise in many industrial situations, such as in blending and reaction vessels for chemical processes. Maintaining the correct levels of liquids in the face of varying demand and other external factors is important in such systems and the development of automatic systems for liquid-level control can be made easier if an appropriate mathematical model is available as a starting point for the control system design. This case study is concerned with the verification and validation of a relatively simple simulation model for a specific two-input two-output system involving two interconnected vessels.

Systems involving fluid flow may be described in a straightforward fashion in mathematical terms, either using distributed parameter models involving partial differential equations, or using lumped parameter models involving ordinary differential equations or differential algebraic equations. The choice depends very much on the intended application and lumped parameter descriptions are widely used in engineering process applications, especially as a basis for controller design.

The interconnected vessels being modelled in this case study form a small laboratory-scale system intended for use in teaching the principles of control engineering. As shown schematically in Fig. 9.1, the system consists of a container (volume 6 l) divided into two tanks of equal dimensions. There are circular holes of various diameters near the base of the partition between the two tanks and the strength of coupling may be changed by inserting or removing plugs as required. The system also has an outflow pipe with a drain tap (under manual control) for adjustment of the output flow rate from one of the tanks. The two tanks have inflows that can be adjusted using electrically-driven variable-speed pumps. The tanks are both equipped with pressure sensors at the base which provide electrical output voltages proportional to the liquid depths.

D.J. Murray-Smith, *Testing and Validation of Computer Simulation Models*,
Simulation Foundations, Methods and Applications,
DOI 10.1007/978-3-319-15099-4_9

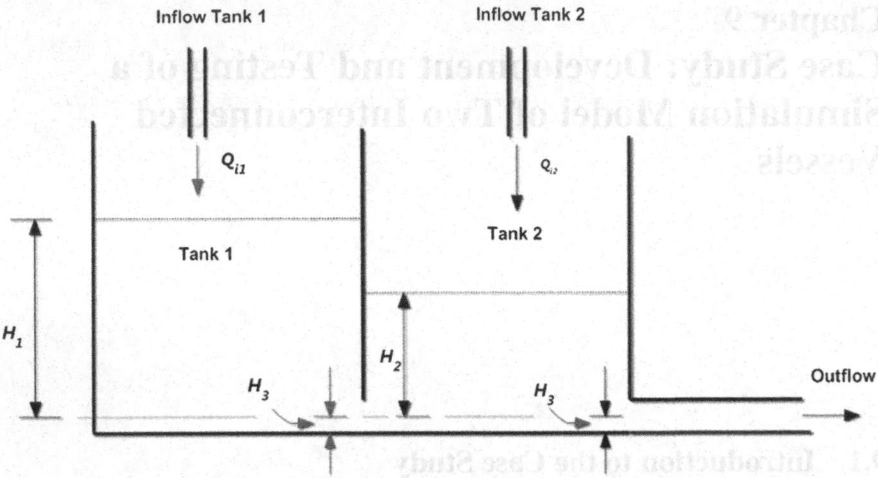

Fig. 9.1 A schematic diagram of the coupled-tanks system. The strength of coupling between the tanks is determined by the areas of holes of circular shape in the wall between the tanks. These holes may be open or plugged. A pipe and drain tap (which is not shown in the diagram) determines the outflow from tank 2

The hardware is based on a commercial product involving a single pump and thus a single flow input (TecQuipment Ltd) [1]. This system, intended primarily for educational use, has been modified at the University of Glasgow through the introduction of a second input involving a second pump and the use of differential-pressure based depth sensors.

The website associated with this book provides a source of additional material relating to this case study. It includes Matlab® files for models of the coupled-tanks system, with detailed documentation, together with files providing some experimental data sets which include relevant information about the test conditions under which the data were collected.

9.2 A Nonlinear Model of the Coupled-Tanks System

A nonlinear mathematical model may be developed for this system using simple and well-known physical principles. The rate of change of volume of liquid in each tank is clearly equal to the difference between the flow rate of liquid into that tank and the flow rate out. Additional background information relevant to the development of this model may be found in the supplementary information through the website associated with this book.

For cases involving liquid levels in Tank 2 that are below the level in Tank 1, the equations, in state-space form, are:

$$\frac{dH_1}{dt} = \frac{Q_{i1}}{A} - \frac{Q_{12}}{A} \tag{9.1}$$

$$\frac{dH_2}{dt} = \frac{Q_{i2}}{A} + \frac{Q_{12}}{A} - \frac{Q_{23}}{A} \tag{9.2}$$

where the flow rates for the external input flows for the two tanks are represented by Q_{i1} and Q_{i2} and H_1 and H_2 are the liquid levels. The cross-sectional area of each tank is represented by the constant A. The flow from tank 1 to tank 2 is represented by Q_{12} and the outflow from the second tank is represented by Q_{23}.

Assuming a perfect thin-walled orifice in the dividing wall between tank 1 and tank 2, it may be shown, from Bernoulli's equation, for conditions involving steady, non-viscous, incompressible flow, that the flow Q_{12} may be related to the liquid levels by expressions that involve a square root function. Thus:

$$Q_{12} = C_{d1} a_1 \sqrt{2g(H_1 - H_2)} \tag{9.3}$$

and similarly, assuming that the outlet from tank 2 also acts as a perfect orifice, it is possible to write the expression for the outflow from the system as:

$$Q_{23} = C_{d2} a_2 \sqrt{2gH_2} \tag{9.4}$$

As shown in Fig. 9.1 the variables H_1 and H_2 are the levels of liquid in each tank above the level of the centre-line of the interconnecting holes and the centre-line of the outlet (H_3). The constant a_1 is the total cross sectional area of the circular holes between the two tanks (termed orifice 1) and a_2 is the cross-sectional area of the outlet pipe from the second tank, leading to the drain tap (termed orifice 2). It should be noted that (9.3) and (9.4) are strictly valid only for ideal orifice flow. The constants C_{d1} and C_{d2} are the discharge coefficients for orifice 1 and orifice 2 respectively. The quantity g in the equations is the gravitational constant.

The key approximations in this model relate to the simple expressions for the flow between the vessels, as given in (9.3) and the flow at the outlet, as given in (9.4). These equations describe ideal flow from a sharp-edged orifice and an expression of this kind may be a reasonable approximation for the flow between the two tanks. On the other hand, the outlet is far from being a simple orifice and involves a pipe, a drain tap and a further section of pipe which has a right-angled bend. Therefore, the approximation given in (9.4) may not be applicable and this part of the model is particularly important in terms of model testing and validation and may require modification. We return to this topic in Sect. 9.5 in the context of model validation and in Sect. 9.6 in terms of model improvement. A similar pair of nonlinear equations may be derived for situations involving liquid levels in Tank 2 that are greater than in Tank 1.

Parameters of the model are defined below and numerical values are shown for the specific hardware system being considered:

Cross-sectional areas of tank 1 and tank 2 $A = 9.7 \times 10^{-3}$ m^2.
Cross-sectional area of orifice 1 $a_1 = 3.956 \times 10^{-5}$ m^2 (using only the two smallest holes).
Cross-sectional area of orifice 2 $a_2 = 3.85 \times 10^{-5}$ m^2 (adjustable from 0 to $a_2 = 3.8 \times 10^{-5}$ m^2).
Height of outlet above base of tank $H_3 = 0.03$ m.
Gravitational constant $g = 9.81$ ms^{-2}.
Maximum flow rate $= Q_{i1\max} = Q_{i2\max} = 5 \times 10^{-5}$ m^{3}s^{-1}.
Maximum liquid level $H_{1\max} = H_{2\max} = 0.27$ m.

The values of the coefficients of discharge C_{d1} and C_{d2} are not known precisely and have to be estimated through experimentation. Methods for the estimation of these two parameters from experimental measurements are discussed in Sects. 9.3.1, 9.3.2, and 9.3.3.

Additional first-order linear ordinary differential equations may be used to describe the dynamic characteristics of the pumps. Using a simple and approximate sub-model for this part of the system, the flow rates $Q_{i1}(t)$ and $Q_{i2}(t)$ may be related to the electrical voltages at the pump inputs, $v_{e1}(t)$ and $v_{e1}(t)$ respectively, by the equations:

$$\frac{dQ_{i1}}{dt} = -\frac{G_{p1}}{T_1} Q_{i1} + \frac{G_{p1}}{T_1} v_{e1} \tag{9.5}$$

$$\frac{dQ_{i2}}{dt} = -\frac{G_{p2}}{T_2} Q_{i1} + \frac{G_{p2}}{T_2} v_{e2} \tag{9.6}$$

The value of the pump calibration constants, G_{p1} and G_{p2}, have been found by experiment to be 7.2×10^{-6} m^{3}s^{-1}V^{-1} and the time constants T_1 and T_2 both have estimated values of 1 s.

Measurement errors in terms of the liquid depth values for data logged using 12-bit analogue-to-digital converters (ADCs) are of the order of ±0.002 m and arise from possible errors in calibrating the depth sensors, together with errors arising from the limited resolution of the converters and the effects of possible external disturbances caused by vibration. Data for steady-state operating conditions are obtained by direct observation of liquid levels, without the use of any ADC, and have errors of the order of ±0.001 m. Experience has shown that the measurement noise introduced by vibrations, can generally be avoided when making steady-state observations as these effects are generally of short duration and are usually associated with movements of people within the laboratory.

9.3 Experiments for Estimation of Parameters C_{d1} and C_{d2}

The form of the system and the fact that we can readily make changes to its structure by closing orifice 1 or orifice 2 during the testing process makes it possible to isolate the effects of the two discharge coefficients and this avoids the need to estimate these two coefficients simultaneously [1]. Using knowledge of the structure of a system to obtain estimates of quantities in this way is often possible and can provide a more effective approach to parameter estimation than analysis of tests on a complete system in its normal mode of operation.

9.3.1 A Test for Estimation of the Discharge Coefficient C_{d1}

In this test the external input to tank 2 is set to zero and the input to tank 1 is used to establish an appropriate initial condition in which the flow into tank 1 is exactly balanced by the flow out of tank 2 with the drain tap fully open [1]. For this balanced situation the liquid levels in the two tanks reach steady values. The inflow to tank 1 is then stopped and simultaneously the drain tap at the outlet of tank 2 is closed (at time $t = t_0$). It follows from (9.1) (9.2), (9.3), and (9.4) that, from that time, the system can be described by the equations:

$$-C_{d1}a_1\sqrt{2g(H_1 - H_2)} = A\frac{dH_1}{dt} \tag{9.7}$$

$$C_{d1}a_1\sqrt{2g(H_1 - H_2)} = A\frac{dH_2}{dt} \tag{9.8}$$

Subtracting (9.8) from (9.7), we find that:

$$-2C_{d1}a_1\sqrt{2g(H_1 - H_2)} = A\frac{d}{dt}(H_1 - H_2) \tag{9.9}$$

Defining $H_\Delta = H_1 - H_2$, (9.2) may be rewritten as:

$$\frac{dH_\Delta}{dt} = -\frac{2C_{d1}a_1\sqrt{2g}}{A}\sqrt{H_\Delta} \tag{9.10}$$

so that

$$-2C_{d1}a_1\frac{\sqrt{2g}}{A}\int_{t_0}^{t_0+T} dt = \int_{H_\Delta(t_0)}^{H_\Delta(t_0+T)} \frac{1}{\sqrt{H_\Delta}}dH_\Delta \tag{9.11}$$

which gives

$$C_{d1} = \frac{A}{a_1 T \sqrt{2g}} \left(\sqrt{H_1(t_0) - H_2(t_0)} - \sqrt{H_1(t_0+T) - H_2(t_0+T)} \right) \quad (9.12)$$

where $H_I(t_O), H_2(t_O), H_I(t_O+T)$ and $H_2(t_O+T)$ are measured quantities at times $t = t_O$ and $t = t_O+T$. The parameter T which defines the time-interval between measurements must be chosen by the experimenter.

One of the difficulties encountered in applying this technique is that it depends critically on the differences between the levels of liquid in the two tanks, both for the initial condition at time t_0 and at time $t_0 + T$. In order to avoid errors in the estimation of the parameter C_{dI} care must be taken to ensure that these differences are not too small and this requires appropriate choices of the initial conditions and also the time interval T over which measurements are made. Since the levels in the two tanks tend to converge, this time interval must be chosen with care.

For operations in the lowest part of the depth range (typically 0.04–0.07 m) a time interval of 6 s was found to be appropriate, while in the upper part (typically 0.175–0.26 m) a time interval of about 13 s was used. Values of C_{dI} were found to vary quite significantly over the range of levels considered, with an estimate of 0.53 for operating conditions in the lowest part of the range, 0.59 in the mid part of the range and 0.63 in the uppermost part of the range. The average value was found to be 0.58.

9.3.2 A Test for Estimation of the Discharge Coefficient C_{d2}

The condition for this test involves opening up all the holes between the two tanks so that they behave, as closely as possible, as a single large tank [1]. With the drain tap fully closed the tanks are filled to an appropriate initial level for the test. We then turn off the pumps providing the external inputs to the tanks and open the drain tap fully. The flow balance condition applied to the two tanks for this condition gives:

$$-C_{d1} a_1 \sqrt{2g(H_1 - H_2)} = A \frac{dH_1}{dt} \quad (9.13)$$

$$C_{d1} a_1 \sqrt{2g(H_1 - H_2)} - C_{d2} a_2 \sqrt{2gH_2} = A \frac{dH_2}{dt} \quad (9.14)$$

Adding (9.13) and (9.14) gives

$$-C_{d2} a_2 \sqrt{2gH_2} = A \left(\frac{dH_1}{dt} + \frac{dH_2}{dt} \right) \quad (9.15)$$

Assuming also that the rates of change of level in the two tanks are approximately equal under these conditions (starting from an initial condition where the initial levels in the two tanks are the same) it follows that:

$$-C_{d2}a_2\sqrt{2gH_2} = 2A\frac{dH_2}{dt} \tag{9.16}$$

Re-arranging and integrating (9.16) gives:

$$C_{d2} = \frac{4A}{a_2 T\sqrt{2g}}\left(\sqrt{H_2(t_0)} - \sqrt{H_2(t_0+T)}\right) \tag{9.17}$$

where $H_2(t_0)$ and $H_2(t_0+T)$ are measured quantities at times $t = t_0$ and $t = t_0+T$. The parameter T, which defines the time-interval between measurements, must be chosen by the experimenter. For H_2 with an initial value of 0.24 m, and plotting $H_2(t)$ versus time (t) it is possible to determine values of H_2 for any number of different values of T. For the purposes of this case study several different values of T were used in the range 45–95 s. The estimates of C_{d2} were found to lie between 0.58 and 0.63, giving an average estimated value of 0.6.

9.3.3 *Estimation of Coefficients C_{d1} and C_{d2} from Test Results for Steady-State Conditions*

Under steady-state conditions with no external input flow to the second tank ($Q_{i2} = 0$) the rate of change of liquid level in each tank must be zero, so that:

$$\frac{dH_1}{dt} = \frac{dH_2}{dt} = 0 \tag{9.18}$$

and, from (9.1) (9.2), (9.3), and (9.4), it follows that:

$$Q_{i1} - C_{d1}a_1\sqrt{2g(H_1 - H_2)} = 0 \tag{9.19}$$

$$C_{d1}a_1\sqrt{2g(H_1 - H_2)} - C_{d2}a_2\sqrt{2gH_2} = 0 \tag{9.20}$$

giving:

$$C_{d1} = \frac{Q_{i1}}{a_1\sqrt{2g(H_1 - H_2)}} \tag{9.21}$$

and

$$C_{d2} = \frac{Q_{i1}}{a_2\sqrt{2gH_2}} \tag{9.22}$$

In the application of this method three cases have been considered with input flows to tank 1 of $1.65 \times 10^{-5}\ m^3s^{-1}$, $2.50 \times 10^{-5}\ m^3s^{-1}$ and $3.33 \times 10^{-5}\ m^3s^{-1}$. The steady-state values of H_1 and H_2 were measured in each of the three cases and the estimated values of C_{d1} were found to be 0.70, 0.68 and 0.71 respectively and the average value of C_{d1} for the three sets of measurements is 0.697. This is larger than the values found using the method of Sect. 9.3.1, which were in the range 0.53–0.63. The corresponding estimates found for C_{d2} were 1.22, 0.67 and 0.57 which involves a much larger range of values than found by the method of Sect. 9.3.2.

The reason for the high value of C_{d2} found in the case of the smallest input flow rate ($1.65 \times 10^{-5}\ m^3s^{-1}$) is likely to be due to the fact that this input flow condition gives a steady state level in tank 2 which is only 0.006 m above the centre-line of the outflow pipe. As mentioned in Sect. 9.1, in the case of the outflow pipe there are grounds to question the assumption of flow through a sharp-edged orifice, upon which the expression for the output flow rate in the model depends. With levels in tank 2 close to the minimum the assumption of orifice flow is of doubtful validity and the very large value found for the discharge coefficient in this part of the operating range provides evidence that reinforces that doubt. Either the structure of the model should be reconsidered to allow for cases of this kind or the operating range over which the existing model may be used needs to be restricted to avoid this region. Using the average of the other two values gives an estimate for C_{d2} of 0.62, which is similar to that found using the method of Sect. 9.3.2 but it should be noted that the time required to obtain the steady state for each value of input flow can involve many minutes and is considerably greater than the time required for the experiments of Sects. 9.3.1 and 9.3.2. This could be viewed as a disadvantage for the steady-state approach.

9.4 Internal Verification of the Simulation Model

Whatever programming language or simulation tools are used, development of a simulation model of the coupled-tanks system is a relatively straightforward process for the mathematical model based on (9.1), (9.2), (9.3), (9.4) and (9.5). Despite that simplicity, verification of the simulation model is still a very important step.

9.4.1 Systematic Checking of Code or Block Diagram Interconnections

The first stage of verification is concerned with checking that the structure of the simulation program is consistent with the form of the mathematical model. This involves working backwards from the statements in the program, on a line-by-line basis, to ensure that when translated into the form of differential and algebraic equations these are the same as those of the original mathematical description. For block-oriented simulation tools, equivalent checks can be carried out starting from the simulation diagram as implemented on the computer and working back to the equivalent equations and comparing those with the mathematical model. Checks should also be made of the parameter values used in the program, or in any input file containing the parameter values, to ensure that the values being used in the simulation correspond exactly to the parameter set of the model itself.

9.4.2 Algorithmic Checks

The second stage of verification is concerned with the choice of algorithms and the overall numerical accuracy. In this application the main issues under this heading are the choice of integration method and issues relating to the communication interval which determines the time interval between output samples (e.g. for display of graphical results). In the case of fixed-step integration methods, comparisons can be made of results obtained with a number of different sizes of integration step and with different integration algorithms. This provides the user with some understanding of the overall suitability of the numerical methods chosen and of the sensitivity of results to the choice of step length. In the case of variable-step integration algorithms, tests can be performed to compare results with different settings of the relative and absolute error limits and with different values of the minimum permitted integration step size. With both the fixed and variable-step methods, some comparisons can also be made using a number of different values of the communication interval to ensure that interesting events in the graphical output from the simulation model are not hidden from the user.

9.4.3 Checks Involving Comparisons of Simulation Results with Analytical Solutions

Once the checks of the structure of the program and of the parameter set being used have been completed, and no algorithmic issues have come to light, a further check can be made by comparing values found from the simulation model with results obtained by analytical methods. Since the mathematical model for the coupled tanks system is nonlinear in form no general solution is available using standard analytical methods. However, as shown in Sect. 9.3.3, it is possible to find steady state solutions for this nonlinear model. Values of liquid levels obtained from the simulation can then be compared with equivalent values found analytically from the mathematical model.

As shown in Sect. 9.3.3, steady-state conditions with no external input flow to the second tank ($Q_{i2} = 0$) result in the equations:

$$\frac{Q_{i1}}{A} - \frac{C_{d1}a_1}{A}\sqrt{2g(H_1 - H_2)} = 0 \tag{9.23}$$

$$\frac{C_{d1}a_1}{A}\sqrt{2g(H_1 - H_2)} - \frac{C_{d2}a_2}{A}\sqrt{2gH_2} = 0 \tag{9.24}$$

Solving these simultaneous equations for the liquid levels gives the following expressions for the steady-state values:

$$H_{1ss} = \frac{Q_{i1}^2}{2g}\left(\frac{1}{C_{d1}^2 a_1^2} + \frac{1}{C_{d2}^2 a_2^2}\right) \tag{9.25}$$

$$H_{2ss} = \frac{Q_{i1}^2}{2gC_{d1}^2 a_1^2} \tag{9.26}$$

If analytical results obtained from (9.25) to (9.26) are not in agreement with simulation results for this special case it is clear that there is an undetected error in the structure or operation of the simulation program. This process should then be repeated for the case involving zero input flow to tank 1 and an appropriate (non-zero) value chosen for the flow into tank 2.

9.5 Validation of the Simulation Model

As has already been emphasised in earlier chapters of this book, there is no single approach to the checking of a mathematical model that can provide the basis for a definitive statement about the overall validity of a model and statements about model validity must always be made in the context of an application. In the case of this coupled-tanks system the simulation model was intended for use in the design of an automatic control system (see e.g. [2, 3]). The specification for the control system required that set levels of liquid would be maintained in both of the tanks in the face of changes of demand in terms of the outflow from the second tank (as determined by the position of the manually-operated drain tap). The form of transient responses in the liquid levels, following demanded changes of level in one or both tanks, should also satisfy given requirements in terms of rise-times and overshoots.

In view of the intended application, it was decided that the validation experiments should be chosen to take account of the likely operating conditions of the system with liquid level control applied. Although the validation tests were all carried out under open-loop conditions without any control loops operating, conditions were chosen to be representative of those in the eventual control application.

It is of course also possible to devise other types of validation test for this system involving more extreme conditions. These could include tests carried out with all the holes in the inter-tank partition blocked off, or tests involving sudden closure of the manually-operated drain tap in the outlet from the second tank. Although of interest and potential importance for some applications, tests of that kind would not necessarily be of immediate importance for the initial design of the liquid-level control system as they may involve physical changes to the structure of the system which are not representative of the intended operating regime.

Experiments carried out on the system for the purposes of model validation had to be designed for input flow values ranging between zero and the maximum permitted value (5×10^{-5} m^3s^{-1}) and also involved conditions involving liquid levels in the two tanks which varied between approximately 0.05 m and the upper limit of 0.27 m (which is the level beyond which overflow occurs). It is important to note that the configuration of the system means that the inputs available for control purposes (the two external input flow rates) can never take negative values and this constraint must also apply to tests carried out on the simulation model. Two types of test were considered. The first of these involved relatively small perturbations of the input flow rates (e.g. about $\pm 0.1 \times 10^{-5}$ m^3s^{-1} or $\pm 0.2 \times 10^{-5}$ m^3s^{-1}) about a selected operating point and gave corresponding small changes in terms of values for H_1 and H_2. The second type of test involved much larger variations in flow rates and therefore much larger changes in the levels.

Steady-state information from measurements and from the simulation model may provide some limited insight, but if this approach is used, it is important not to base any comparisons on steady state results that may have been used in the estimation of parameters. Although it appears to be a very simple approach, experience with the use of this method suggests that it provide much less information for model validation purposes than dynamic tests involving comparisons of time-history data. Also, as noted in the discussion in Sect. 9.3.3, the time required for steady-state testing can be considerable since sufficient time must be allowed for transient effects to have become negligible before the measurements can be taken.

9.5.1 Graphical Methods

Appropriate qualitative methods that could be used in an application such as this, with only two output variables, are likely to involve direct comparisons of experimental and predicted time histories. Separate plots of the time histories of the measured and model variables may be augmented by plots of differences between system and model variables and by quantitative measures of model performance, such as Theil's Inequality Coefficient, appropriate fitness functions, or simple statistical measures such as those provided by box plots.

Figures 9.2 and 9.3 show typical sets of time-domain data that allow comparisons of measured levels with corresponding simulation results. In this particular case the model parameter values are as given in Sect. 9.2 above and values used for

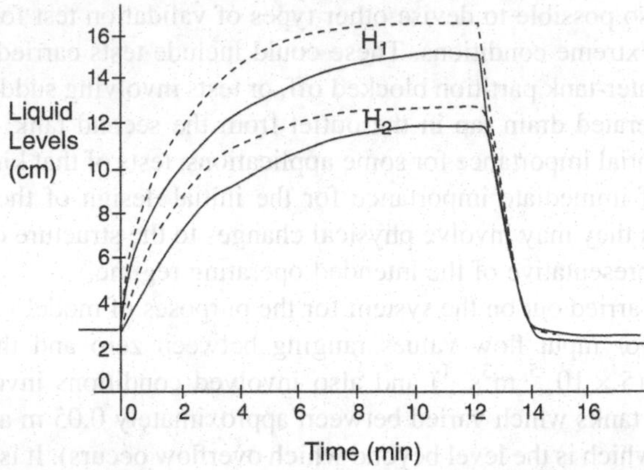

Fig. 9.2 Simulation results for H_1 and H_2 (*dashed lines*) and the corresponding measured responses from the coupled-tanks system (*continuous lines*) for an experiment with input flow rate of zero to tank 2 and an input involving step changes of input flow rate to tank 1 from zero to 2.67×10^{-5} m^3/s and then back to zero, with initial levels of 0.03 m, measured from the base of the tanks (From Ref. [2])

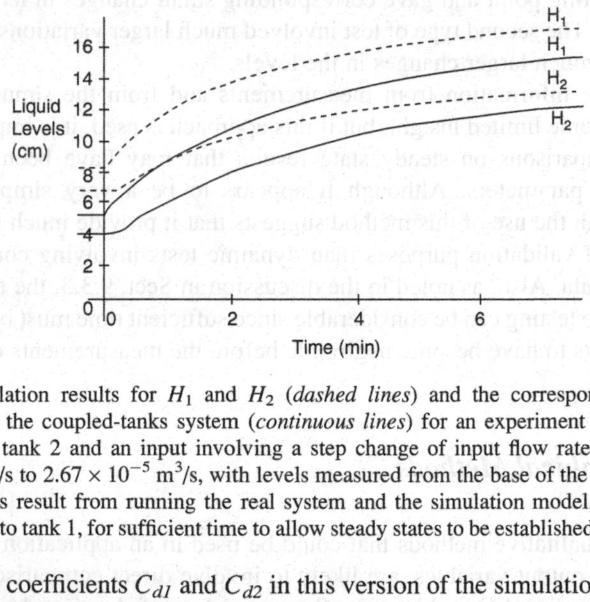

Fig. 9.3 Simulation results for H_1 and H_2 (*dashed lines*) and the corresponding measured responses from the coupled-tanks system (*continuous lines*) for an experiment with input flow rate of zero to tank 2 and an input involving a step change of input flow rate to tank 1 from 1.67×10^{-5} m^3/s to 2.67×10^{-5} m^3/s, with levels measured from the base of the tanks. Note that the initial levels result from running the real system and the simulation model, with the given initial flow rate to tank 1, for sufficient time to allow steady states to be established (From Ref. [2])

the discharge coefficients C_{d1} and C_{d2} in this version of the simulation model were 0.75 and 0.5 respectively. In Fig. 9.2 the applied test input case involved zero flow to tank 2 and a change of input flow to tank 1 from zero to 2.67×10^{-5} m^3s^{-1} and then back to zero after about 13 min. The results of Fig. 9.3 involve different initial conditions and a shorter experiment, although the pattern of input flow variation involved the same magnitude of step change as in the case shown in Fig. 9.2. Note that in Figs. 9.2 and 9.3 the liquid levels shown differ slightly from previous results since the values of H_1 and H_2 in this case are measured from the base of the tanks and therefore include the value H_3. The results of both these transient tests show that significant differences exist between the measured responses and the responses predicted from the simulation. One feature of interest in the results shown in Fig. 9.2 is the highly unsymmetrical nature of the positive-going and negative-going transients. It is clear that the quality of fit between measurements and the simulation model results is poor and that changes in the model structure or in the values of the C_{d1} and C_{d2} parameters should be considered.

9.5.2 Application of the Model Distortion Approach

The model distortion method outlined in Sect. 7.2.4 has been applied with success to try to establish the extent to which the discharge coefficients have to be varied in order to obtain a match between measured and simulated responses [4]. The test conditions considered in that investigation involved a simple input in the form of a step change in external input flow rate to tank 1 from 1.72×10^{-5} m^3s^{-1} to 2.56×10^{-5} m^3s^{-1} applied at time $t = 50$ s for 450 s. The external flow into tank 2 was zero throughout and the initial conditions corresponded to $H_1 = 0.026$ m and $H_2 = 0.01$ m, meaning that the two tanks were initially almost empty. This initial condition was chosen deliberately in order to include the part of the operating range that had produced large estimates of the discharge coefficient C_{d2} using the steady state method of parameter estimation. Following the applied change of input flow at time $t = 50$ s the liquid levels in the two tanks increased slowly and the measured values of the measured liquid levels at time $t = 500$ s were found to be $H_1 = 0.1$ m and $H_2 = 0.065$ m.

Application of the model distortion method, as described in Chap. 7, resulted in the time history plots shown in Figs. 9.4 and 9.5 for the values of the discharge coefficients C_{d1} and C_{d2} needed to get an exact match between the simulated and experimental responses. These time history plots for the discharge coefficient parameters are noisy but show clearly that the value of C_{d1} is approximately constant and has an average of about 0.85, while the value for C_{d2} varies to a greater extent, being about 0.78 during the initial steady state when the tanks are almost empty and then falling steadily to about 0.67 at the end of the test period.

It is interesting to compare these results, obtained using the distortion method, with the corresponding values estimated in Sects. 9.3.1, 9.3.2, and 9.3.3. The value of C_{d1} of about 0.85 is larger than the value of 0.697 found through the steady-state method of Sect. 9.3.3 but it is significant that this value found by distortion does not vary over the period of the experiment.

The value of 0.67 for C_{d2} at the end of the model distortion test agrees with the result found using the steady-state method of parameter estimation described in Sect. 9.3.3 for the steady-state involving the middle flow condition which gave levels of liquid in tanks 1 and 2 of 0.085 m and 0.041 m respectively. The initial higher level of C_{d2} found using model distortion is consistent with the high value for this parameter estimated using the steady-state method for the lowest input flow condition which gave a steady state condition in which the liquid levels were close to their minimum values.

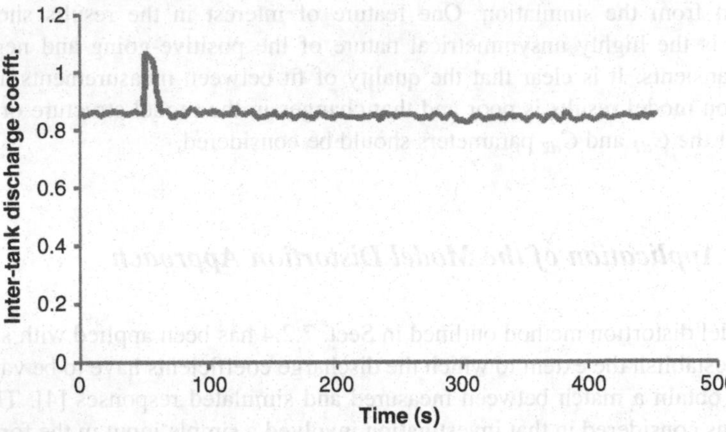

Fig. 9.4 Results from application of the model distortion method: calculated time variation of the coefficient C_{d1} to achieve fit of model output to measured level in tank 2

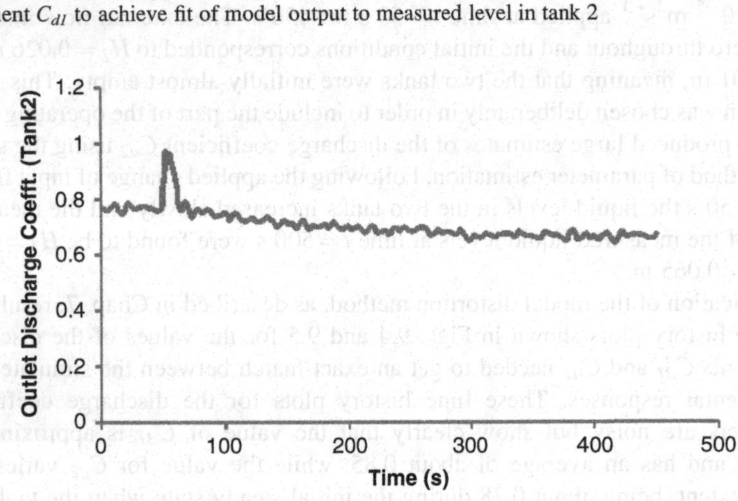

Fig. 9.5 Results from application of the model distortion method: calculated time variation of the coefficient C_{d2} to achieve fit of model output to measured level in tank 2

Results found using the model distortion method were checked by running the simulation with the measured input and with values of C_{d1} and C_{d2} which varied with time according to the time histories of Figs. 9.4 and 9.5. As should be expected, the resulting time histories of the two liquid levels obtained from the simulation matched the measured time histories for those variables exactly.

9.5.3 Application of Genetic Programming (GP) Techniques

For application of the GP approach to nonlinear system identification, outlined in Chap. 5, the experiments considered were similar in form to those used with the model distortion method. Only the flow input to tank 1 was varied and the input flow to tank 2 was maintained at zero throughout. A binary form of input signal was used as pump input, with amplitude and frequency chosen to cover the full operating range of liquid levels and an appropriate frequency range. Two sets of data were collected, one of which was used for system identification and the other for model validation. The test input signals for these two cases were broadly similar but did not involve exactly the same amplitude and frequency distributions.

The GP method was used to develop an algebraic expression for the flow out of the second tank. This expression included parameters estimated using a combined simplex and simulated-annealing method [5].

The first step was based on an approach which involved solution of nonlinear flow equations at each point and storing the resulting information for laminar and turbulent flow conditions in two look-up tables giving the flow rate for every possible value of the liquid level H_2 [5]. The laminar and turbulent flow look-up tables then formed two of the non-terminal nodes in the tree structure for the application of the GP algorithm. Other non-terminal nodes were for addition, subtraction, multiplication by a gain factor, a square root function, a unit time delay and a basis function. The basis function was introduced to allow the GP to evolve different dynamics for the output flow depending on the liquid depth in tank 2. This involved a Gaussian function which had characteristics that depended on H_2 [5]. The terminal nodes were the system input (the external flow rate into tank 1), the state variables H_1 and H_2 and the numerical values zero and one. At each iteration of the GP software an expression for the output flow was incorporated into a nonlinear simulation model of the two tank system (in place of the standard expression in (9.2) and (9.4) involving the discharge coefficient C_{d2}). The optimisation process was based on evaluation of the sum of the squares of the differences between predicted values of H_2 from the simulation model and the corresponding measured values. Since the evolutionary processes involved in the GP algorithm incorporate random components, as outlined in Chap. 5, the complete optimisation procedure was repeated four times and the resulting forms for the flow equation from tank 2 were compared. One of the four solutions involved a much larger value of the sum of the squares of the errors than the other three, so a decision was taken to discard this result. A common feature of the other three solutions was that they were dominated by terms involving the output of the laminar flow look-up table multiplied by one basis function and one of these was selected as the best representation [5]. The measured response data (continuous line) and the model predictions for the model identified through use of the GP algorithm (dash-dot line) are shown in Fig. 9.6 along with the response for the model of (9.1), (9.2), (9.3), and (9.4) using the best estimate of the discharge coefficients (dashed line). The results show that both the conventional model and the model derived using the GP

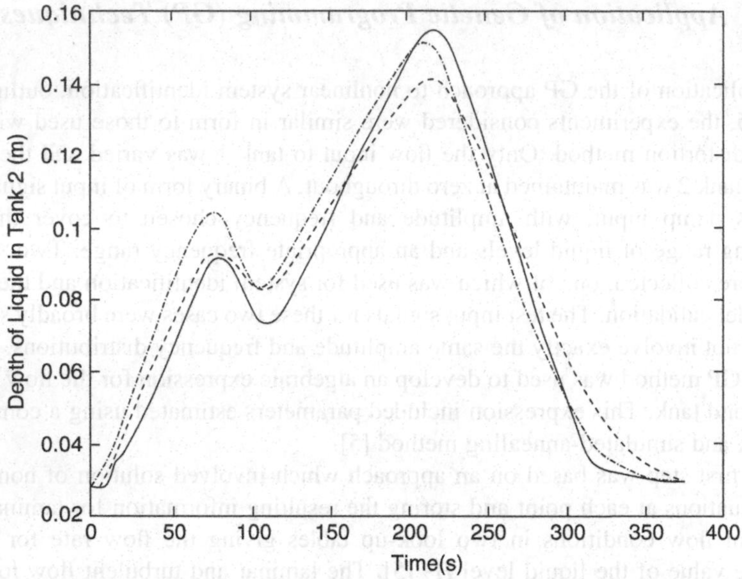

Fig. 9.6 Comparison of results from application of GP approach (*dash-dot curve*) with measured response data for depth of liquid in tank2 (*continuous curve*) and conventional model based on estimated values of discharge coefficients (*dashed curve*). The horizontal axis in this diagram is from time $t = 0$ to time $t = 460$ s. The vertical axis, representing the depth of liquid measured from the base of the tanks, has a range from 0.02 to 0.16 m (From Ref. [5])

algorithm give a reasonable fit for the variable H_2. Both models were then subjected to a different form of test input to provide a basis for model validation. The results are shown in Fig. 9.7 and indicate that the GP-derived model gives a much better fit to the measured response than the conventional model based on the orifice flow representation for the outlet from tank 2. The error in the prediction for the case of orifice flow increases significantly with time over the period of the experiment.

9.5.4 Application of the Inverse Simulation Approach

Inverse simulation techniques have also been considered as an approach to validation of the nonlinear coupled-tanks model [6]. The test input applied to generate the measured response data was similar to that applied using the model distortion approach and involved the application of step changes of flow to tank 1. Applying the measured time history of the level in tank 1 as input to an inverse simulation model gave a time history of predicted input flow that showed some distinctive differences compared with the ideal step input. In particular, the input calculated from the inverse simulation model was found to be greater than the actual input over the periods of time when the forward simulation model gave an output smaller

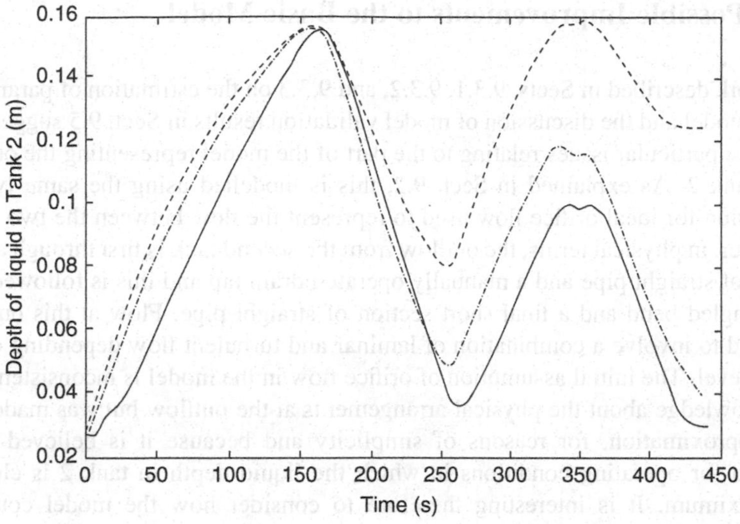

Fig. 9.7 Comparison of results for tests of the coupled-tanks system models using data sets not applied in the identification of the model using the GP approach. The results for the GP-derived model (*dash-dot curve*) can be compared with measured response data for the depth of liquid in tank 2 (*continuous curve*) and also with results for the conventional model based on estimated values of discharge coefficients (*dashed curve*). The horizontal axis in this diagram is from time $t = 0$ to time $t = 460$ s. The vertical axis, representing the depth of liquid measured from the base of the tanks, has a range from 0.02 to 0.16 m (From Ref. [5])

than the measured response. Conversely the input calculated from inverse simulation was slightly below the true input for periods when the predicted H_1 value from the forward model was above the measured response [6]. These findings are not surprising and information of this kind, expressed in quantitative form, could clearly provide a basis for accepting or rejecting the model. However, it has to be recognised that this information from the inverse simulation results did not, in this particular case, provide any new physical insight or additional clues about aspects of the model that could be changed to give an improved fit. Parameter sensitivity analysis could perhaps be helpful in analysing possible sources of error in the model but it was concluded that, for this specific application, the inverse simulation approach did not provide information that was significantly different from that available from validation methods based on conventional forward simulation. It should be noted that this conclusion should not be taken to suggest that inverse simulation methods are of no potential value for simulation model validation in general and a different conclusion about the benefits of these methods might well be reached in another type of application.

9.6 Possible Improvements to the Basic Model

The work described in Sects. 9.3.1, 9.3.2, and 9.3.3 on the estimation of parameters of the model and the discussion of model validation results in Sect. 9.5 suggest that there are particular issues relating to the part of the model representing the outflow from tank 2. As explained in Sect. 9.2, this is modelled using the same type of expression for ideal orifice flow used to represent the flow between the two tanks. However, in physical terms, the outflow from the second tank is first through a short length of straight pipe and a manually operated drain tap and this is followed by a right-angled bend and a final short section of straight pipe. Flow at this outlet is believed to involve a combination of laminar and turbulent flow depending on the liquid level. The initial assumption of orifice flow in the model is inconsistent with this knowledge about the physical arrangements at the outflow but was made, as a first approximation, for reasons of simplicity and because it is believed to be justified for operating conditions in which the liquid depth in tank 2 is close to the maximum. It is interesting therefore to consider how the model could be improved.

Attempts to adjust the values of the discharge coefficients C_{d1} and C_{d2} have shown that there is no single value for each of the parameters that gives a good match to experimental results over the whole of the operating range of the system and this finding is supported by the results obtained using the model distortion approach. Another approach is to use estimated values of the discharge coefficients at a number of different operating points to form an empirical model. Although not based on physical laws and principles, such a model has been shown to allow predictions to be made of level variations in the two tanks over the full operating range of the system which are more accurate than those provided by other models. An example of this type of modification, involving a fixed value of C_{d1} and a variable value of C_{d2}, based on estimated values for 12 operating points has been discussed by Murray-Smith and Gong [2].

The results obtained using the Genetic Programming methodology for nonlinear system identification provide an immediate indication of possible structural changes in the model and, through repeated optimisation processes, allow comparisons to be made between different candidate models. In the results discussed in Sect. 9.5.3, it was suggested that an improved the representation could be found for the flow at the outlet of tank 2 using a combination of look-up tables developed from solution of nonlinear fluid-flow equations and an appropriate basis function. Such a modification could avoid the assumption of ideal orifice flow at the outlet from tank 2 and extends the range of operating conditions over which the model could be used with confidence.

References

1. Wellstead PE (1981) Coupled tanks apparatus, manual. TecQuipment Ltd., Nottingham
2. Gong M, Murray-Smith DJ (1998) A practical exercise in simulation model validation. Math Comput Model Dyn Syst 4(1):100–117
3. Murray-Smith DJ (2012) An application of the individual channel analysis and design approach to control of a two-input two-output coupled-tanks system. Acta Polytech 52(4):121–134
4. Gray GJ, Lew KV, Murray-Smith DJ (1997) Applications of the distortion method for model validation. In: Troch I, Breitenecker F (eds) Proceedings 2nd MATHMOD VIENNA IMACS symposium on mathematical modelling, February 1997. Argesim, Vienna, pp 1033–1038
5. Gray GJ, Murray-Smith DJ, Li Y et al (1998) Nonlinear model structure identification using genetic programming. Control Eng Pract 6:1341–1352
6. Murray-Smith DJ, Wong BO (1997) Inverse simulation techniques applied to the external validation of nonlinear dynamic models. In: Luker P (ed) UKSim'97, third conference of the United Kingdom simulation society, Keswick, April 1997. UKSim, Edinburgh, pp 100–104

Chapter 10
Case Study: Model Validation and Experiment Design for Helicopter Simulation Model Development and Applications

Helicopters and other forms of rotorcraft present special challenges for modelling, simulation and flight control system design. Flight testing of the vehicle is also difficult because, for some flight conditions, most helicopters have unstable modes and this can severely limit the duration of flight experiments. Experimental results may also be adversely affected by high vibration levels in flight and therefore sensor noise tends to be a problem for in-flight measurements in general and for model validation and for the application of system identification and parameter estimation techniques in particular.

Active control technology (the "fly-by-wire" approach) began to be applied to helicopters in the late 1980s and 1990s. By that time this approach to flight control was already well established for civil and military fixed-wing aircraft. It was clear that the improvement in capabilities expected from this active-control technology was achievable only if accurate and proven mathematical models were available (see, e.g. [1]). The publication of revised handling qualities requirements for military helicopters in the USA [2] provided a timely stimulus to these developments.

10.1 Problem Areas in Helicopter Modelling and Simulation

Although improved simulation models have been developed and control system design techniques which exploit the multi-input multi-output structure of helicopters have been applied with some success, accuracy limitations remain a problem with models of these vehicles and this can still seriously limit flight control system performance. In this field of application it is especially important that models have appropriate accuracy over a defined range of frequencies and for a range of manoeuvre amplitudes. For example, high-bandwidth model-following flight

D.J. Murray-Smith, *Testing and Validation of Computer Simulation Models*,
Simulation Foundations, Methods and Applications,
DOI 10.1007/978-3-319-15099-4_10

control systems often have feed-forward control pathways to provide improved agility for large and rapid manoeuvres but this approach can only succeed with accurate models of the vehicle (see, e.g. [1, 3]). System identification can also be important for the validation of ground-based simulators for rotorcraft and highly accurate mathematical models are needed for simulators for pilot training or for use in helicopter research and development (see, e.g. [1]).

Simulation techniques using nonlinear, physically-based, models are potentially very important for investigating performance limitations and for flight control system design and are used throughout the development process for a new vehicle, from requirements specification and initial design stages to eventual certification. Typical uses include:

- Developing flight control laws for given requirements in terms of handling qualities and disturbance rejection characteristics.
- Establishing whether or not control and stability characteristics of the vehicle are likely to be satisfactory, especially towards the limits of the flight envelope.
- Investigating effects of failures of major components of the vehicle, such as an engine failure, in different phases of flight and the associated piloting strategies to ensure safe recovery.

Model validation is a central issue since the accuracy of model predictions is vitally important for all these applications. The range of operations within which the performance is adequate for the intended application is important, both in terms of frequencies (reflecting the modal characteristics) and amplitudes (reflecting the nonlinear behaviour) and both have to be investigated [4]. However, the analysis of flight test data for model validation is complicated by the high vibration environment within helicopters and, even with appropriate pre-processing and filtering of data, the problems encountered are often more challenging than in many other application areas in engineering. However, better models offer many potential benefits and are essential for the design of high-bandwidth active flight control systems. It is now widely accepted that models used for flight control design should include coupled body/rotor dynamics and, in this respect, theoretical models with parameters derived from wind-tunnel investigations are of limited value. Experimental modelling, using system identification and parameter estimation methods, is now being recognised as being increasingly important for the estimation of poorly defined parameters and for model validation.

10.1.1 Quantitative Criteria for Functional Validation of Helicopter Models

There are several aspects of performance where broad agreement should be expected between vehicle behaviour and the predictions of an accepted helicopter simulation model. These include:

- Predictions of *trim states* in which steady control settings ensure that the vehicle maintains a specified steady-state flight condition, such as straight and level flight with a specified forward airspeed. Predicted trim states should match those found from flight tests to within a specified percentage of the full trim range for the control inputs, for attitude angles and in terms of the power required.
- Predictions of stability characteristics should match estimates obtained from flight data to within a certain percentage of the modulus of the eigenvalue concerned.
- In general terms, predicted responses from the simulation model should match responses from flight to within a certain percentage of full range at a specified time after the application of the input. The issue of time is important in this type of application because after 20 or 30 s small differences in initial conditions in the model and system may show up as much larger differences in responses.

Some authors have provided specific criteria that may be useful for assessing helicopter simulation models. Examples can be found in the work of Padfield and Du Val [4], Tischler and Remple [5], and Lu et al. [6]. Useful information is also contained within the NATO-supported AGARD (Advisory Group on Aerospace Research and Development) Advisory Report 280 prepared by members of the AGARD Flight Mechanics Panel Working Group WG18 [7]. In some cases these criteria relate to time-domain measures, while in others frequency-domain measures are emphasised. Both these classes of measure are important, with time-domain measures often being given particular emphasis in work relating to helicopter handling qualities investigations. Because of the significance of frequency-domain techniques for control system design, measures that involve frequency-response information are often emphasised in the case of models being applied in flight control system design.

10.1.2 Investigation of Physical Fidelity in Helicopter Simulation Models

Physical fidelity relates to the quality of the model in terms of the accuracy of the underlying modelling assumptions. Overall goodness-of-fit is clearly relevant but, for many applications, it is also important to consider whether or not the model structure is appropriate for the conditions for which the model is to be used. For example, a linear model for a particular operating point within the flight envelope may be very useful for the initial stages of flight control system design (where linear control system design techniques are applied) but a more detailed nonlinear model would be needed in order to carry out simulated tests of the flight control system for large manoeuvres. Similarly, the inclusion of actuator limits, and especially actuator rate limits, may be vital for any model that is used for investigation of flight control and handling qualities issues. There are many other issues where physical fidelity is important, including situations where interactions between fuselage dynamics and rotor dynamics are known to be significant. Questions relating to the margin of control available to the pilot in any specific manoeuvre are particularly important.

Any simplifications used in the representation need to be considered carefully in the light of the available flight test data and the intended application of the model. Residual differences between variables measured in flight and the corresponding model variables can often provide important clues about inadequacies in terms of the assumed model structure (see, e.g. [7]).

10.2 System Identification and Parameter Estimation in Helicopter Modelling

System identification and parameter estimation techniques can be used not only for the estimation of uncertain parameters but also for validation of models developed using physical laws and principles. Taken alone, simulation models derived on a physical basis and using wind–tunnel data are seldom found to be adequate for flight-control system design. Such model deficiencies are most often not apparent until the flight testing of a prototype vehicle, leading to costly redesign, extended flight test programmes with a modified prototype vehicle and delays in certification. With modern methods of control system design, such as the H^{∞} approach, flight control systems can often be made robust to compensate for model errors but this can often be achieved only at the expense of some aspects of performance.

The benefits of system identification, as seen from within the aircraft industry, appear to relate mainly to a reduction in flight testing hours needed for certification of new designs and for fine tuning of the vehicle's agility and handling qualities. Identification methods may also be useful for improving engineers' confidence in physically-based models used in design and for reducing levels of uncertainty. Estimation of parameters from flight is now an increasingly important part of prototype testing and is especially relevant for some aerodynamic stability and control parameters. Although flight testing is costly, additional tests carried out specifically for system identification and parameter estimation purposes could allow essential design modifications to be made through virtual prototyping methods more quickly, more efficiently, and at lower cost than by using traditional approaches. Industry views on the potential benefits of system identification and parameter estimation techniques may be found in the results of a survey by Padfield published in the early 1990s, as discussed in Chap. 8. From published work and contacts made with industry, since the date of that survey, it appears unlikely that industry views have changed significantly.

Early published accounts describing the application of system identification techniques to helicopters mainly involved time-domain methods. However, since the late 1980s, the use of frequency-domain methods has been seen to have some advantages. In order to use frequency-domain methods measurements have to be transformed into the frequency domain. This can be done using the Fast Fourier Transformation or the Chirp z-Transform [7–9]. The frequency-domain approach has the advantage of allowing particular parts of the frequency range to be emphasised, with data lying outside the parts of the range of interest being given less attention or discarded. Thus, for example, for the estimation of parameters of a six-degrees-of-freedom rigid-body model, rotor degrees of freedom can sometimes be excluded as they involve higher frequencies for some types of helicopter. Similarly, if the objective is the identification of rotor dynamics, the lower part of the frequency range which is associated with the rigid-body dynamics may be excluded. This can be viewed as a form of model reduction within the framework of system identification and parameter estimation [10]. Frequency-domain methods have been applied successfully for a number of projects involving both rigid-body models and rotor models and were also used extensively by the NATO-supported AGARD Flight Mechanics Panel Working Group WG18 for work leading to the preparation of the AGARD Advisory Report 280 on Rotorcraft System Identification [7].

Figure 10.1 shows records of flight test data, as measured in a flight experiment on a BO105 helicopter [7], together with simulation results generated using a well-established helicopter flight mechanics model involving use of the Deutsches Zentrum fűr Luft- und Raumfahrt (DLR – German aerospace centre) simulation program SIMH [11]. The uppermost traces show test input time histories for the lateral and longitudinal cyclic controls. The records immediately below are for roll-rate and pitch-rate variables predicted from the SIMH simulation model (dashed lines) together with the corresponding variables measured in the flight test programme (continuous lines). Significant differences between the simulation results and the measured results are immediately evident and suggest that this physically-based simulation model may have some significant limitations. The two sets of results below these show the same flight test measurements along with outputs predicted using a model based on parameters estimated from flight tests. Differences between the measured responses and the predictions from the simulation model are much smaller in this case, suggesting that the model based on parameter estimates is a more accurate representation of the vehicle.

Although the results obtained using the identified model and shown in the lower traces of Fig. 10.1 may be regarded as encouraging, it has to be recognised that helicopters present some special difficulties when identification methods are applied. For example, as mentioned earlier in this chapter, it is normal to be faced with short test records. In many cases the duration may even be short compared with the dominant time constants and periods of the dominant oscillatory modes. Also, in the multi-input multi-output case especially, these models involve large numbers of parameters and a wide range of frequencies that may be of interest. As indicated in Sect. 10.1, measurements can also involve high levels of sensor noise and this is another undesirable feature for system identification and usually necessitates a significant amount of effort in terms of pre-processing of flight test data. Information about some pre-processing techniques that are applied in practice may

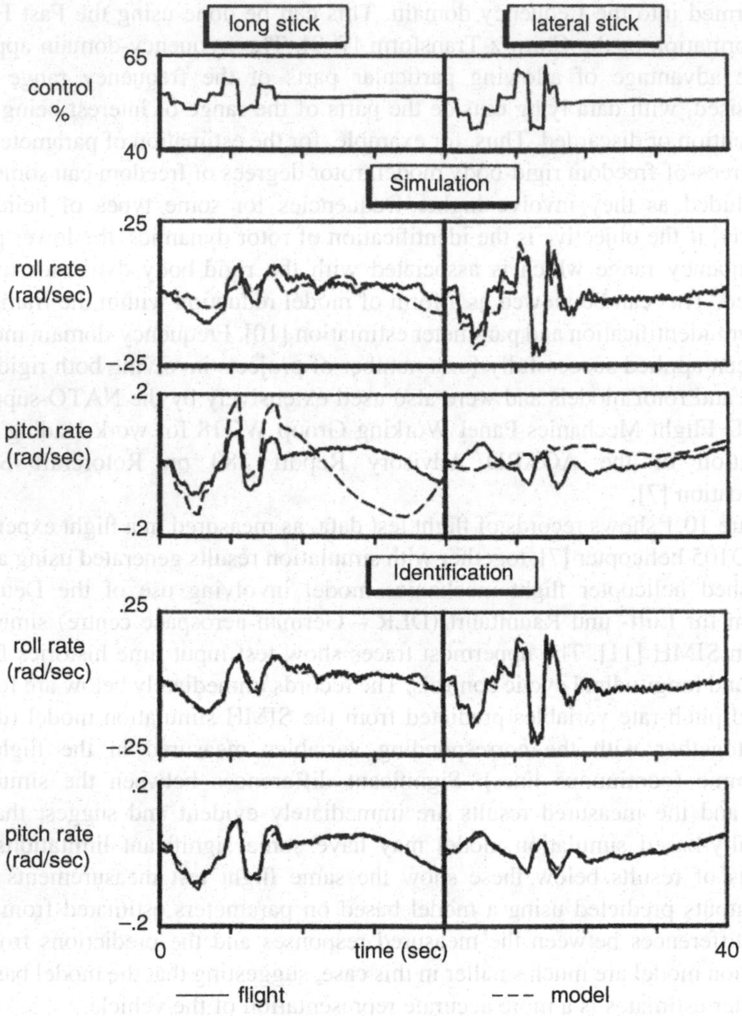

Fig. 10.1 The top three time-domain records show roll rate and pitch rate responses for longitudinal and lateral test inputs from flight test results from a Bo-105 helicopter together with simulation results found for the same test inputs using the DLR SIMH simulation model. The results shown in the two bottom records are for a simulation involving an identified linear model. The period of each experimental record is approximately 24 s. In each of the records simulation results are indicated by *dashed lines* and the measured output responses are shown by *continuous curves*. Measured data have been subjected to pre-processing to check compatibility of the signals from different sensors, to correct subsequently for measurement errors and to evaluate the data quality. The original version of this figure was published by the Advisory Group for Aerospace Research and Development, North Atlantic Treaty Organisation (AGARD/NATO) in AGARD Advisory Report 280 'Rotorcraft System Identification', September 1991 [7]

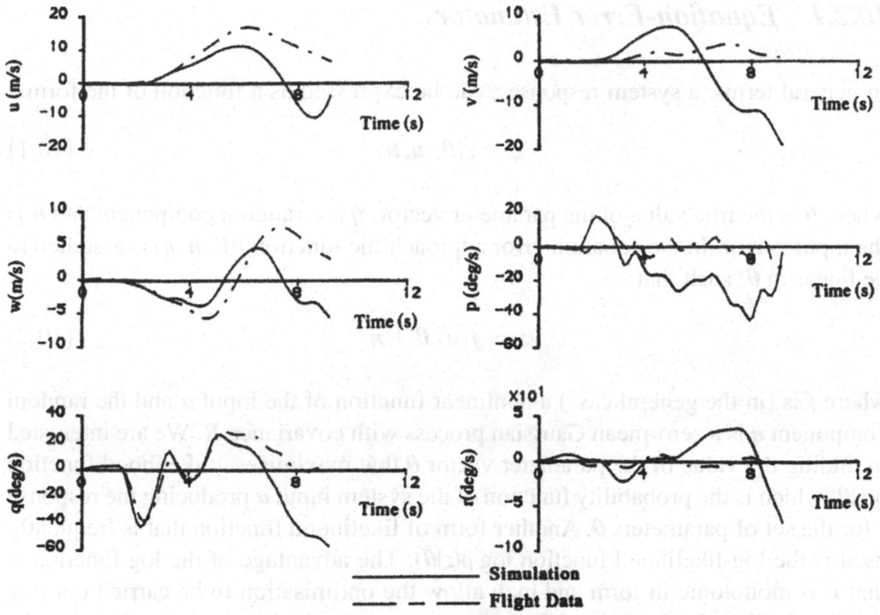

Fig. 10.2 Simulation data and flight data compared for the same test inputs for a Westland Lynx helicopter. Simulation predictions are shown by *continuous lines* and flight data are shown by *dashed lines*. Measured data have been subjected to pre-processing to check compatibility of the signals from different sensors, to correct subsequently for measurement errors and to evaluate the data quality. From the *top left* the results show responses in terms of forward velocity (*left*), lateral velocity (*right*), vertical velocity (*left*), roll rate (*right*), pitch rate (*left*) and yaw rate (*right*) (From [12])

be found in the AGARD Advisory Report 280 [7] and also in the book by Tischler and Remple [9].

Figure 10.2 further illustrates the problem of relatively poor predictions from physically-based simulation models and also demonstrates some of the problems associated with short test records. The data source in this case is a flight test carried out on a Westland Lynx helicopter [12] and the results show that, for the specific manoeuvres considered, differences between measurements and predictions are significant in some cases and tend to increase with time. The simulation results are from a specific physically-based nonlinear simulation model.

Theoretical properties of different types of estimator must be considered carefully in any application [13]. The methods of system identification most widely used in helicopter flight testing can be categorised broadly into "equation-error" methods and "output-error" methods. If certain conditions are met (which depend on the characteristics of the models involved) output error methods can be shown to act as "maximum-likelihood" estimators.

10.2.1 Equation-Error Estimators

In general terms, a system response z can be expressed as a function of the form:

$$z = z(\boldsymbol{\theta}, \boldsymbol{u}, \boldsymbol{\eta}) \tag{10.1}$$

where $\boldsymbol{\theta}$ is the true value of the parameter vector, $\boldsymbol{\eta}$ is a random component and $\boldsymbol{u}$ is the input vector. In the equation-error approach the function $f(\boldsymbol{\theta}, \boldsymbol{u}, \boldsymbol{\eta})$ is assumed to be linear in $\boldsymbol{\theta}$, such that

$$z = f(\boldsymbol{u}).\boldsymbol{\theta} + \boldsymbol{\eta} \tag{10.2}$$

where f is (in the general case) a nonlinear function of the input $\boldsymbol{u}$ and the random component $\boldsymbol{\eta}$ is a zero-mean Gaussian process with covariance $\boldsymbol{R}$. We are interested in finding the value of the parameter vector θ that maximises a likelihood function $p(z|\boldsymbol{\theta})$ which is the probability function of the system input $\boldsymbol{u}$ producing the response z for the set of parameters $\boldsymbol{\theta}$. Another form of likelihood function that is frequently used is the log-likelihood function $\log p(z|\boldsymbol{\theta})$. The advantage of the log function is that it is monotonic in form and may allow the optimisation to be carried out in a simpler fashion. In the case of the equation-error method it can be shown that the log-likelihood function has the form [14]:

$$\log p(z|\boldsymbol{\theta}) = -\frac{1}{2}(z - f(\boldsymbol{u})\boldsymbol{\theta})^T \boldsymbol{R}^{-1}(z - f(\boldsymbol{u})\boldsymbol{\theta}) - \frac{1}{2}\log(2\pi\boldsymbol{R}) \tag{10.3}$$

This is maximised if we can find the condition for the minimum of the term:

$$(z - f(\boldsymbol{u})\boldsymbol{\theta})^T \boldsymbol{R}^{-1}(z - f(\boldsymbol{u})\boldsymbol{\theta}) \tag{10.4}$$

which is the same as the square of the error between the measured variable z and the model response, which is represented by $f(\boldsymbol{u})\boldsymbol{\theta}$. The equation-error estimator is therefore a form of least-squares estimation algorithm.

The expression (10.4) can be minimised using an analytical approach [14] to give:

$$\boldsymbol{\theta} = \left(f(\boldsymbol{u})^T \boldsymbol{R}^{-1} f(\boldsymbol{u})\right)^{-1} \left(f(\boldsymbol{u})^T \boldsymbol{R}^{-1} z\right) \tag{10.5}$$

Consider a helicopter model in linear state-space form:

$$\dot{\boldsymbol{x}}(t) = \boldsymbol{A}\boldsymbol{x}(t) + \boldsymbol{B}\boldsymbol{u}(t) \tag{10.6}$$

$$\boldsymbol{y}(t) = \boldsymbol{C}\boldsymbol{x}(t) + \boldsymbol{\eta}(t) \tag{10.7}$$

where $\boldsymbol{x}(t)$ is the vector of state variables for the model, $\boldsymbol{u}(t)$ is the vector of model inputs, $\boldsymbol{y}(t)$ represents the model outputs (corresponding to the measured responses used in system identification), $\boldsymbol{\eta}(t)$ represents Gaussian noise components with zero mean and covariance $\boldsymbol{R}$ and $\boldsymbol{A}$, $\boldsymbol{B}$ and $\boldsymbol{C}$ are the model state, input and output matrices.

This model can be re-organised in the form:

$$\dot{\boldsymbol{x}}(t) = \begin{bmatrix} \boldsymbol{A} & 0 \\ 0 & \boldsymbol{B} \end{bmatrix} \begin{bmatrix} \boldsymbol{x}(t) \\ \boldsymbol{u}(t) \end{bmatrix} \tag{10.8}$$

which, in the time domain, is equivalent to an equation:

$$\boldsymbol{z}^*(t) = \boldsymbol{F}\boldsymbol{u}^*(t) \tag{10.9}$$

where $\boldsymbol{z}^*(t) = \dot{\boldsymbol{x}}(t), \boldsymbol{u}^*(t) = \begin{bmatrix} \boldsymbol{x}(t) \\ \boldsymbol{u}(t) \end{bmatrix}$ and $\boldsymbol{F} = \begin{bmatrix} \boldsymbol{A} & 0 \\ 0 & \boldsymbol{B} \end{bmatrix}$

This allows the equation-error approach to be used to obtain estimates of the elements of the matrix $\boldsymbol{F}$ provided measurements of $\boldsymbol{x}(t)$, $\boldsymbol{u}$ and $\dot{\boldsymbol{x}}$ are available. In general, both $\boldsymbol{u}^*(t)$ and $z^*(t)$ include sensor noise and this presents a potential difficulty in that the form of equation defining the equation-error approach (i.e. (10.9)) is based on an assumption that the input does not contain noise. This assumption is reasonable since input measurements normally contain less noise than other measured variables. However, significant numerical difficulties can arise with this method if the matrix $\boldsymbol{f}(\boldsymbol{u})^T \boldsymbol{R}^{-1} \boldsymbol{f}(\boldsymbol{u})$ is ill-conditioned [15], since this matrix must be inverted to obtain the parameter vector $\boldsymbol{\theta}$. This ill-conditioning is usually caused by correlation between the elements of the input vector $\boldsymbol{u}$. The best way of avoiding such problems is through the test input design process.

10.2.2 Output-Error Estimators

In the output-error approach the response function z is the vector of output variables for a linear state-space model of the form:

$$\dot{\boldsymbol{x}}(t) = \boldsymbol{A}\boldsymbol{x}(t) + \boldsymbol{B}\boldsymbol{u}(t) \tag{10.10}$$

$$\boldsymbol{z}(\boldsymbol{\theta}, \boldsymbol{u}, \boldsymbol{\eta}, t) = \boldsymbol{C}\boldsymbol{x}(t) + \boldsymbol{\eta}(t) \tag{10.11}$$

The matrices $\boldsymbol{A}$, $\boldsymbol{B}$ and $\boldsymbol{C}$ are functions of θ. The variable $\boldsymbol{\eta}(t)$ is Gaussian with zero mean value and covariance $\boldsymbol{R}$ and $\boldsymbol{u}(t)$ is the vector of inputs.

It may be shown that, in this case, the log-likelihood function is [14]:

$$\log \mathrm{p}(z|\boldsymbol{\theta}) = -\frac{1}{2}(z(t) - \boldsymbol{C}\boldsymbol{x}(t))^T \boldsymbol{R}^{-1}(z(t) - \boldsymbol{C}\boldsymbol{x}(t)) - \frac{1}{2}\log(2\pi \boldsymbol{R}) \tag{10.12}$$

For any implementation of the method the output variable $z(t)$ becomes a vector of measured quantities which have to be compared with sampled values of the model outputs at the same time instants. There is no simple analytical solution in terms of maximising the log-likelihood function since the model response is not linear in the unknown parameters. Numerical methods must therefore be used to find the conditions for the maximum [14]. This approach requires measurements of the outputs $z(t)$ and the inputs $\boldsymbol{u}(t)$ but, unlike the equation-error method, not the complete set of state variables $\boldsymbol{x}(t)$ or the derivatives $\dot{\boldsymbol{x}}(t)$.

In general, the output-error approach may be expected to provide more accurate estimates than the equation-error method. In numerical terms, however, the output-error approach is significantly more demanding and the algorithm may not converge if initial values of unknown parameters are too far from the optimum values. Since the equation-error approach is much simpler to apply it is often used as an initial step to find approximate and preliminary parameter estimates, prior to the use of these as initial estimates in an output-error algorithm (see e.g. [16]).

10.2.3 Maximum-Likelihood Estimators

The maximum likelihood approach [14] is a specific type of output-error identification method which may allow parameter estimates to be found for linearised aircraft models from flight data involving both measurement and process noise. The latter could include, for example, effects of un-modelled atmospheric disturbances.

The maximum-likelihood estimate is defined as the value of the parameter vector $\boldsymbol{\theta}$ that maximises the log-likelihood function $\log \mathrm{p}(z|\boldsymbol{\theta})$. Important properties of these estimators include the following:

- Estimates are asymptotically unbiased, meaning that with an infinite number of measurements there is zero bias on estimates.
- Estimates are asymptotically efficient, meaning that the estimates have the minimum possible covariance (as given by the Cramer-Rao bound [17]).

As with other implementations of the output-error type of approach, the maximum-likelihood approach is an iterative method but, unlike some other output-error techniques it involves estimation of noise statistics through determination of the measurement noise covariance matrix.

The implementation of the maximum-likelihood algorithm places additional demands on the processor being used and there may be problems of non-convergence, depending on the initial values used for the components of the parameter vector. Thus, like other output-error techniques, the maximum likelihood algorithm is used mostly as a second stage in the identification process. Initial investigations using an equation error approach provide appropriate parameter values to start the second stage using a maximum-likelihood or other output-error algorithm. In practice most published results of helicopter system identification testing have involved the use of output-error methods but it is fair to say that relatively few cases have been based on a full implementation of the classical maximum-likelihood algorithm.

It is important to note that the assumptions implicit in all the identification approaches considered here of no state noise and random measurement noise of a simple type may not apply in practice. For example, the measurement errors may well include modelling errors because of limited knowledge of the dynamic behaviour of the air data system [7]. Turbulence effects and model uncertainties may also produce noise on the state variables. With these non-ideal situations and other uncertainties in terms of the model structure it is clear that the use of output-error techniques do not necessarily give unbiased estimates of the model parameters. However, the measures of quality of estimates such as those given by the Cramer-Rao bounds, do still provide an indication, in a relative sense, of the sub-set of parameters that are most successfully estimated and also of parameters that may be less important.

10.3 Test Input Design for Helicopter System Identification and Model Validation

Test inputs have been found to be of considerable importance for helicopter system identification. They are equally important for validation of helicopter simulation models. Figure 10.3 shows idealised forms of two commonly used test inputs which have been widely used for many years for helicopter system identification and model validation. One of these signals is the doublet which has a particularly simple binary form involving a positive going step from some initial constant value, followed by a short period of constant input and then a negative-going step of twice the magnitude of the original positive step. After a further period of constant negative input there is a positive-going step back to the original input value. In Fig. 10.3a, which shows the doublet signal, it is assumed that the input is initially zero. The time periods of the constant positive and negative inputs are the same and this period is chosen as part of the experimental design, as is the initial constant level of input prior to the doublet perturbation and the amplitude of the doublet signal itself. The second type of signal is the 3-2-1-1 input and an idealised signal of this kind, shown in Fig. 10.3b, is similar to the doublet but involves an initial positive step input for a period of three time units, followed by negative step to a new negative constant value which is held for two time units, a further positive input for one time unit, followed by a second negative step and then a return to the starting value after a total of seven time units. Binary multi-step signals of this kind have the advantage that they can be applied relatively easily through the pilot's controls, although achieving the required amplitudes of input and the precise timing required for the transitions does require considerable skill.

Another form of input that has received considerable attention in recent years is the "frequency-sweep" signal which is an approximately sinusoidal signal of steadily increasing frequency. This, again, can be applied manually through the pilot's controls, and like other manually-applied signals, requires considerable skill

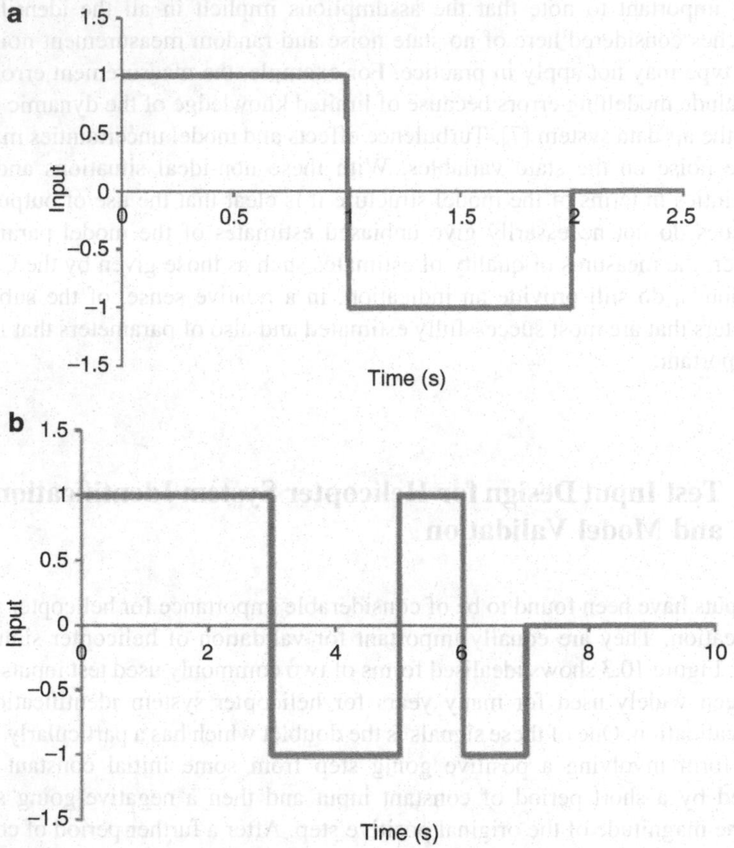

Fig. 10.3 Idealised forms of test signal (**a**) a doublet signal and (**b**) a 3-2-1-1 signal. It should be noted that the initial values of signals are zero at time $t = 0$

in its successful application. Figure 10.4 shows practical signals, as applied to a helicopter in one specific set of flight tests. In the case of the doublet and 3-2-1-1 inputs these differ slightly from the idealised versions shown above, especially in the case of the 3-2-1-1 where the amplitude varies during the initial stage and the second stage is clearly less than two time units in duration. The frequency sweep signal includes a broader range of frequencies than the other two inputs, although the record length is also greater than for the 3-2-1-1 or doublet. This can be seen clearly from the roll rate and pitch rate responses to the three inputs. In cases involving tests on a vehicle with small stability margins, or where unstable modes are present, the potential for using long input sequences like the frequency sweep may be limited.

Test input signal design specifically for the identification of a helicopter, as with other systems, inevitably involves the use of a model of the vehicle. As discussed in Chap. 7, uncertainties within that model mean that the resulting flight experiments are unlikely to be optimal. A first step in the identification process should be the

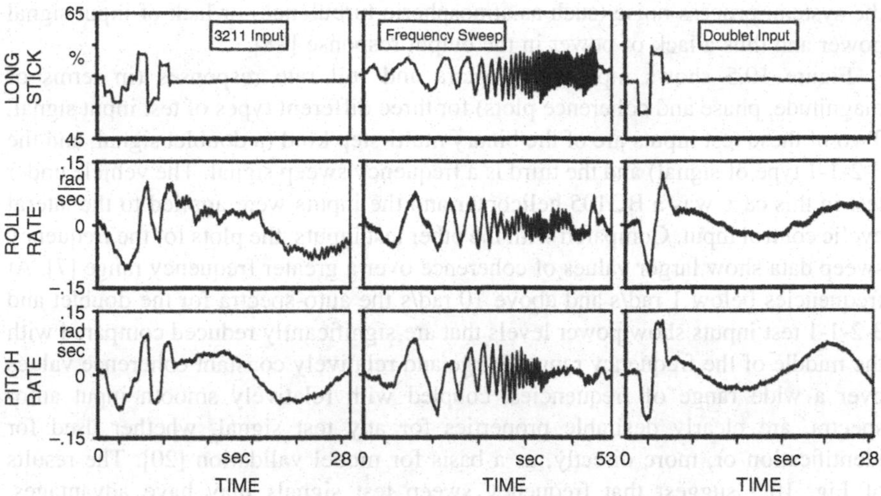

Fig. 10.4 Three commonly used test inputs for system identification and external validation of a helicopter simulation model. Measured responses are also shown for the roll rate and pitch rate variables. In this case the vehicle under test was a BO-105 helicopter. Measured data have been subjected to pre-processing to check compatibility of the signals from different sensors, to correct subsequently for measurement errors and to evaluate the data quality. The original version of this figure was published by the Advisory Group for Aerospace Research and Development, North Atlantic Treaty Organisation (AGARD/NATO) in AGARD Advisory Report 280 'Rotorcraft System Identification', September 1991 [7]

setting up of a relatively simple model of the vehicle that can be of assistance in the design of test inputs for system identification. This initial model may be physically based or experimentally derived using relatively simple flight experiments.

Once an approximate model of the vehicle is available the processes of experiment design can be started. Clearly, we need a quantitative basis for comparing different test signals and, as discussed in Chap. 7, this involves quantities such as the parameter information matrix and the dispersion matrix. It should be noted that, if the experiment design is being carried out in order to apply system identification and parameter estimation techniques, consideration has to be given to the properties of the estimation algorithms involved. For example, inputs designed using measures based on the dispersion matrix have been found to be especially useful in cases where long test records are available, provided asymptotically efficient parameter estimation methods are used, such as the maximum-likelihood approach.

Spectral properties of test inputs are also very important and coherence functions provide a useful measure of the degree to which a given signal provides satisfactory excitation [18, 19]. As discussed in Chap. 7, these functions provide a measure of the fraction of the output auto-spectrum which may be accounted for by a linear relationship with the input auto-spectrum. Ideally the coherence is unity over the frequency range of interest and coherency values significantly smaller than one may indicate a number of potential difficulties. These include nonlinearities within the system, process noise (such as atmospheric turbulence) or lack of input signal power and thus a lack of power in the output response [18].

Figure 10.5 shows input auto-spectra and roll rate responses (in terms of magnitude, phase and coherence plots) for three different types of test input signal. Two of these test inputs are of the binary multi-step kind (a doublet signal, and the 3-2-1-1 type of signal) and the third is a frequency sweep signal. The vehicle under test in this case was a BO105 helicopter and the inputs were applied to the lateral cyclic control input. Compared with the other test inputs, the plots for the frequency sweep data show larger values of coherence over a greater frequency range [7]. At frequencies below 1 rad/s and above 10 rad/s the auto-spectra for the doublet and 3-2-1-1 test inputs show power levels that are significantly reduced compared with the middle of the frequency range. Large and relatively constant coherence values over a wide range of frequencies, coupled with relatively smooth input auto-spectra, are clearly desirable properties for any test signal, whether used for identification or, more directly, as a basis for model validation [20]. The results of Fig. 10.5 suggest that frequency sweep test signals may have advantages, provided practical difficulties do not result from the longer time period required for the application of this type of test signal.

10.3.1 Design for a Specified Input Auto-spectrum

In cases where the aim of the system identification process is validation of linearised flight mechanics models, the inputs used for testing must be consistent with the modelling assumptions and the input design must take full account of any constraints. It is also important to obtain long test records since parameter estimates then have time to converge and efficient estimation (i.e. minimum variance estimation) is possible. Long test records also ensure that the lower frequency oscillatory modes of the vehicle response are well defined in the measured output records and they also allow the formal requirements of criteria based on the dispersion matrix to be satisfied. Research by Leith and Murray-Smith [21] was carried out with the aim of designing a test input that would provide long test records and also an acceptably "small" value for the dispersion matrix. It is important to avoid resonances in the system since an input that excites resonances could rapidly produce a large response that might require the flight experiment to be prematurely aborted. Inputs should also be chosen to ensure that the signal has no steady state component as a constant component in the input will tend to produce a steady-state constant response component which shifts the operating point. If the operating point is significantly different from that used for linearisation of the theoretical model, the parameter estimates found by experiment will not be consistent with the model, thus invalidating the whole procedure.

A method of autospectrum design is required that allows users to choose inputs that are relatively simple in form so that they can be applied manually by the pilot, while also avoiding known resonances, thus ensuring that relatively long test

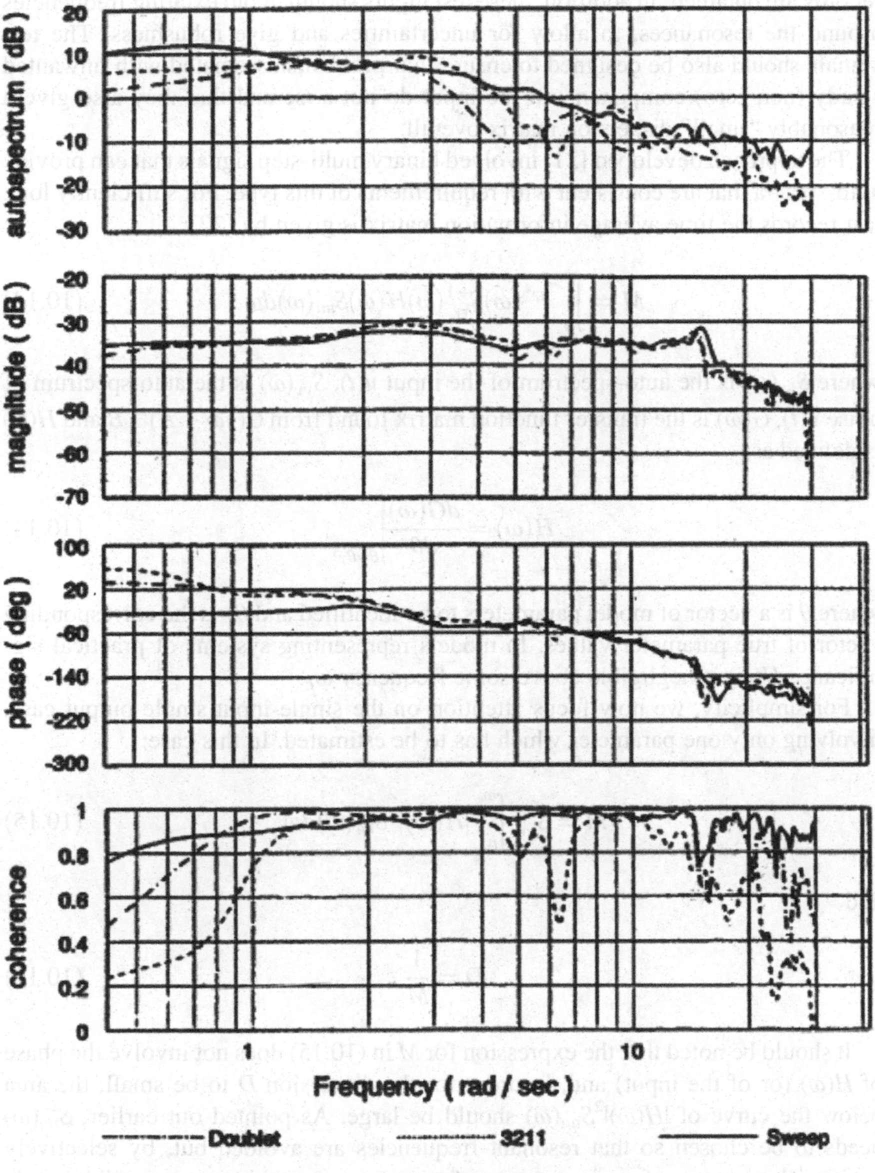

Fig. 10.5 Input auto-spectra and roll rate responses (In terms of magnitude, phase and coherence plots) of a BO 105 helicopter for three types of test input signal (doublet, 3-2-1-1 and frequency sweep). Test inputs were applied to the lateral cyclic control input. The frequency sweep results are shown by the *continuous line* while results from the other two inputs (doublet and 3-2-1-1) are shown by *dash-dot* and *dashed* patterns respectively. Measured data have been subjected to pre-processing to check compatibility of the signals from different sensors, to correct subsequently for measurement errors and to evaluate the data quality. The original version of this figure was published by the Advisory Group for Aerospace Research and Development, North Atlantic Treaty Organisation (AGARD/NATO) in AGARD Advisory Report 280 'Rotorcraft System Identification', September 1991 [7]

records are obtained. In addition, these test inputs should avoid exciting frequencies around the resonances, to allow for uncertainties and give robustness. The test signals should also be designed to ensure that problems associated with unwanted steady (non-zero) components in the input do not arise and that they also give a reasonably "small" dispersion matrix overall.

The approach developed [21] involved binary multi-step signals that can provide auto-spectra that are consistent with requirements of this type. For sufficiently long test records the time average information matrix is given by [22]:

$$M = \int_0^{\infty} F^*(\omega) S_{vv}^{-1}(\omega) H(\omega) S_{uu}(\omega) d\omega \tag{10.13}$$

where $S_{uu}(\omega)$ is the auto-spectrum of the input $u(t)$, $S_{vv}(\omega)$ is the auto-spectrum of noise $\eta(t)$, $G(\omega)$ is the transfer function matrix found from $C(j\omega - A)^{-1}B$ and $H(\omega)$ is defined as:

$$H(\omega) = \left. \frac{dG(\omega)}{d\theta} \right|_{\theta=\theta_T} \tag{10.14}$$

where θ is a vector of model parameters to be identified and θ_T is the corresponding vector of true parameter values. In models representing systems of practical significance $H(\omega)$ is negligible above some frequency ω_c.

For simplicity, we now focus attention on the single-input single-output case, involving only one parameter which has to be estimated. In this case:

$$M = S_{vv}^{-1} \int_0^{\omega_c} |H(\omega)|^2 S_{uu}(\omega) d\omega \tag{10.15}$$

and

$$D = \frac{1}{M} \tag{10.16}$$

It should be noted that the expression for M in (10.15) does not involve the phase of $H(\omega)$ (or of the input) and, in order for the dispersion D to be small, the area below the curve of $|H(\omega)|^2 S_{uu}(\omega)$ should be large. As pointed out earlier, $S_{uu}(\omega)$ needs to be chosen so that resonant frequencies are avoided but, by selectively exciting the system at frequencies away from resonances, this area can still be made large. The same argument applies in the general case where more than one parameter has to be found and where there is more than one input and output variable.

It is important to note that there are many approximations and simplifications in this whole approach to test input design based on the properties of the information matrix and the spectral properties of the model. However, the approach can be justified when it is remembered that the system transfer function matrix $G(\omega)$ is known only approximately at the stage when test inputs are being designed and the quantity $H(\omega)$ is therefore even less well known. Only the very general characteristics of $G(\omega)$ and $H(\omega)$ can be relied upon.

Now, consider the form of a general aperiodic binary multi-step signal. It is possible to show [21] that the Fourier transform of a signal of this type has the form:

$$F(\omega) = \frac{1}{j\omega}\left[1 + 2\sum\nolimits_{i=1}^{n-1} (-1)^i e^{(-j\omega t_i)} + (-1)^n e^{(-j\omega t_n)}\right] \tag{10.17}$$

where $F(\omega)$ is the Fourier transform of the input signal

ω is the frequency (rad/s)
$n+1$ is the number of steps in the input signal
t_i is the time (s) at which the i th step occurs for an initial time $t_0 = 0$.

A cost function J can now be defined such that:

$$J = \sum\nolimits_{k=1}^{m} a_k |F(\omega_k)|^2 \tag{10.18}$$

where a_k are constants and ω_k are frequencies for $k = 1, 2, \ldots\ldots, m$.

An optimal spectrum can now be found based on the number of steps $(n+1)$ in the input, the number of weighting constants (m) and the values of the weights a_k and associated frequencies ω_k in the cost function J to determine the frequencies that should and should not be emphasised in the auto-spectrum of the signal. A large positive coefficient a_k should produce an input having a large component within the auto-spectrum at the frequency ω_k whereas a large negative a_k value gives a small auto-spectrum component at that frequency.

This approach has been applied with considerable success to the design of binary multi-step test inputs for a Westland Lynx helicopter to try to avoid the problems of short test records that can arise with the use of more conventional test inputs. The model considered involved the pitching moment equation involving seven parameters that had to be estimated. The form of the equation is as follows:

$$\frac{dq(t)}{dt} = M_u u(t) + M_w w(t) + M_q q(t) + M_v v(t) + M_p p(t) + M_{\eta 1s}\eta_{1s} + M_{\eta 1c}\eta_{1c} \tag{10.19}$$

where $u(t)$ is the longitudinal velocity, $w(t)$ is the vertical velocity, $q(t)$ is the pitch rate, $v(t)$ is the lateral velocity, $p(t)$ is the roll rate, η_{1s} is the longitudinal cyclic control input and η_{1c} is the lateral cyclic control input. The quantities M_u, M_w, M_q, M_v, M_p, $M_{\eta 1s}$ and $M_{\eta 1c}$ are the parameters to be estimated. The flight condition considered involved 80 knots level flight.

A frequency-domain output-error method [16] was chosen for use in the identification process and the first step in the test signal design process involved finding theoretical impulse responses for the Lynx helicopter for the chosen flight condition using the HELISTAB helicopter flight mechanics model developed at the UK Royal Aerospace Establishment, Bedford [23]. Transformation of the impulse responses to the frequency domain showed the presence of a large peak in the magnitude responses at about 0.3 rad/s, accompanied by a rapid change of phase. This corresponds to a known unstable phugoid mode for the chosen flight condition of 80 knots straight and level flight at a theoretical natural frequency of 0.36 rad/s.

The test signal design process was used to ensure that the input auto-spectrum had no component at zero frequency (i.e. no "dc" component), that it avoided the known resonance at about 0.3 rad/s and that the input test signal was capable of exciting the system at frequencies between 2 and 3 rad/s but not above 3 rad/s. That upper limit was introduced because previous experience suggested that the theoretical model was useful only for frequencies below about 3 rad/s since, at higher frequencies, dynamic effects within the rotor sub-system could have a significant influence and those modes were not included in the model.

The optimisation process led to a binary multi-step input which involved five steps, with an auto-spectrum in which most of the power was concentrated within a narrow band of frequencies between 1 and 3 rad/s. The only difficulty with this input design was that the timings of the transitions between the two levels were irregular and it was considered impractical to require a pilot to apply this signal manually. The input was therefore modified to give simpler timings that would be acceptable to pilots. Since the optimal test signal had been designed for robustness due to model uncertainties it was believed that this slight change of timings would have a small effect. The resulting "double doublet" input is shown in Fig. 10.6a, along with its auto-spectrum in Fig. 10.6b. It can be seen that the auto-spectrum of this simpler signal also shows the highest level of input power at frequencies between 1 and 3 rad/s.

From the flight trials it was found that typical record lengths for the double-doublet input of the order of 30 s could be achieved routinely with a Lynx helicopter. This compared with 10–15 s for a traditional doublet input and only 3 s for the 3-2-1-1 type of input. Estimates of the seven parameters of the pitching moment equation were obtained successfully from the flight data using the frequency-domain equation-error approach [21]. Overall, the double doublet gave system identification results that were consistently better than those obtained from the use of other inputs and this test input was also found to be robust both to slight timing errors in the manual application of the signal and to uncertainties associated with the theoretical model used in its design [21].

It should be noted that the frequency-domain emphasis in the system identification approach being applied in this work and the characterisation of the test signal in frequency domain terms was important in the context of the eventual application of the helicopter model for flight control system design. For that application the helicopter model must perform especially well close to open-loop gain and phase cross-over frequencies and a frequency-domain approach can be helpful in ensuring that this requirement is met, as well as providing useful physical insight. Although the test inputs were intended primarily for use in system identification and parameter estimation in the work reported by Leith and Murray-Smith [21], the criteria

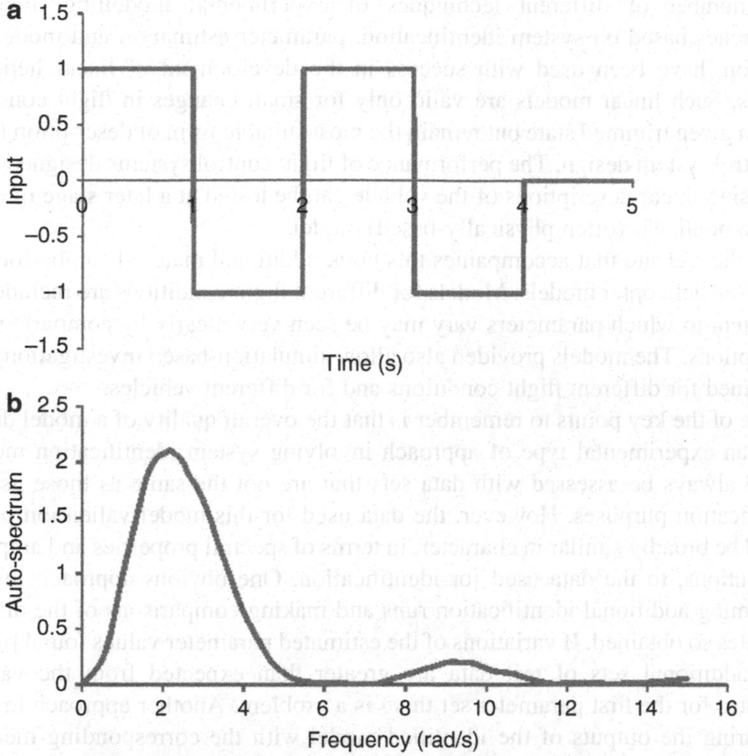

Fig. 10.6 Idealised form of "double-doublet" signal (a) and its autospectrum (b). It should be noted that the initial value of the test signal is zero at time $t = 0$

used in their design are equally relevant for the design of test inputs for model validation purposes.

10.4 The Validation of Six-Degrees-of-Freedom Helicopter Models

Important issues that have to be taken into account in the validation of helicopter simulation models include the frequency range over which model quality should be assessed and the amplitude range for each variable. For example, models intended for use in the design of active (computer-based) flight-control systems are particularly demanding since the frequency range of interest extends well beyond the upper frequency limit for control by a human pilot (5 rad/s approximately) to include a much wider range of frequencies (from zero to about 20 rad/s). The amplitudes to be considered for each of the model variables also depend upon the intended application of the model.

A number of different techniques of experimental modelling, involving approaches based on system identification, parameter estimation and model optimisation, have been used with success in the development of linear helicopter models. Such linear models are valid only for small changes in flight conditions about a given trimmed state but remain the most suitable form of description for use in control system design. The performance of flight control systems designed in this way using linear descriptions of the vehicle can be tested at a later stage of design using a nonlinear (often physically-based) model.

On the website that accompanies this book additional material can be found on linearised helicopter models. Models for different flight conditions are included and the extent to which parameters vary may be seen very clearly by comparing these descriptions. The models provided also allow simulation-based investigations to be performed for different flight conditions and for different vehicles.

One of the key points to remember is that the overall quality of a model derived using an experimental type of approach involving system identification methods should always be assessed with data sets that are not the same as those used for identification purposes. However, the data used for this model validation process should be broadly similar in character, in terms of spectral properties and amplitude distributions, to the data used for identification. One obvious approach involves performing additional identification runs and making comparisons of the different estimates so obtained. If variations of the estimated parameter values found from all these additional sets of test data are greater than expected from the variance estimates for the first parameter set there is a problem. Another approach involves comparing the outputs of the identified model with the corresponding measured responses using the input records from the additional test results. Predicted outputs from the model may then be compared directly with the corresponding measurements using graphical methods, in conjunction (perhaps) with some of the quantitative measures discussed in Chap. 3.

An example of the use of his latter approach is shown in Fig. 10.7 where results are presented of validation tests on an identified model of a SA-330 Puma helicopter. This second (validation) stage of the experimental modelling process is based on a test input that was not applied during the identification process. For validation purposes, output variables from the model were compared directly with the corresponding measurements using a graphical approach. The first and third columns of Fig. 10.7 show the quality of the fit found following the system identification process. For the data sets shown in the first column the input was the lateral cyclic control moved to the left and for the records shown in the third column the input was the lateral cyclic input moved to the right. In this case an almost perfect fit was obtained between flight data and the identified model outputs. The second and fourth columns show the quality of fit found when the identified models were subjected to other forms of test input. The quality of fit in this case is seen to be significantly poorer. The results also show an asymmetry in that the validation results for the lateral left cyclic control input are closer to the ideal (for the three variables considered) than results for the case involving the lateral right cyclic input. This suggests a need for additional tuning of the identified model in terms of

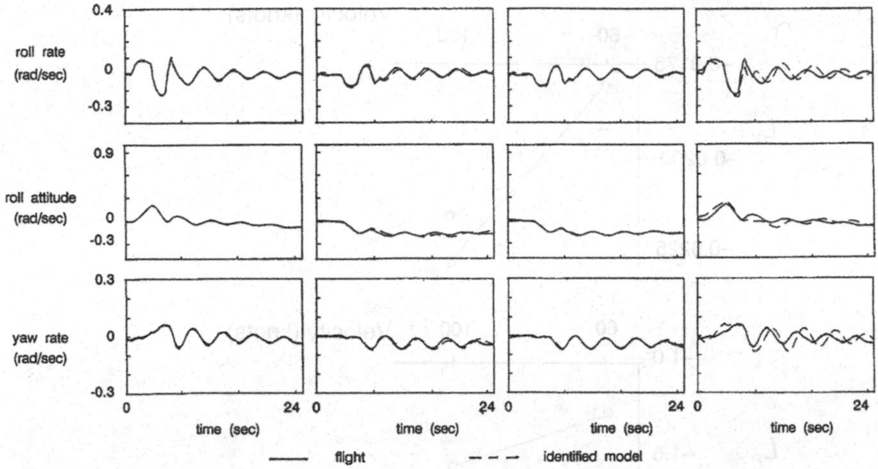

Fig. 10.7 Tests of a model identified from flight data (SA-330 Puma helicopter). The first and third columns of the records show the fit obtained through system identification. In the case of the first column the input was the lateral left cyclic control and for the third column of the records the input was the lateral right cyclic input. The second and fourth columns show the fits obtained when those experimentally-derived models were subjected to a different time history of test input. Measured data have been subjected to pre-processing to check compatibility of the signals from different sensors, to correct subsequently for measurement errors and to evaluate the data quality. The original version of this figure was published by the Advisory Group for Aerospace Research and Development, North Atlantic Treaty Organisation (AGARD/NATO) in AGARD Advisory Report 280 'Rotorcraft System Identification', September 1991 [7]

responses to lateral cyclic inputs. Sensitivity analysis could be used to check for possible parameter dependencies and further sets of flight data could be used to investigate this further. Possibly the most important point about this example is that it shows the importance of splitting available test data into different sets, with some being used for identification and others for the validation process.

Nonlinear helicopter models derived mainly using physical laws and principles may also be validated using data from flight experiments. One approach that has been applied with success involves estimation of key parameters within linearised models of the vehicle, using flight data collected during identification experiments performed for a number of different flight conditions. This allows comparisons to be made of the parameter values estimated from flight data with the equivalent values found by linearisation of the physically-based nonlinear flight mechanics model [17]. Although this approach is valid only for small perturbations from the chosen trimmed flight condition, it can provide insight about the underlying nonlinear model from the comparisons made at different operating points. Consistent changes in the estimated and theoretical values of a parameter following a change in the operating condition (such as the forward speed of the vehicle) tend to enhance the credibility of the nonlinear model. On the other hand, if the trend in theoretical values of a chosen parameter is not the same as the trend in the estimated

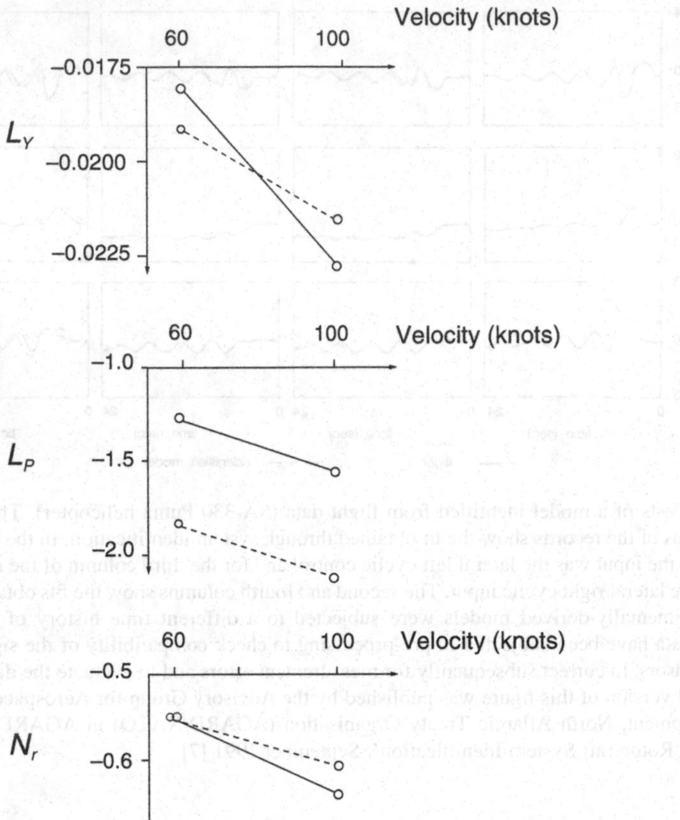

Fig. 10.8 Parametric trends for two flight conditions, showing theoretical results from linearised models derived form a physically-based nonlinear simulation model (HELISTAB) [23] (*dashed lines*) and corresponding parameter estimates from flight tests and the use of system identification methods (*continuous lines*) (From [12])

values for that parameter, possible changes in the nonlinear model should be considered. This is illustrated in Fig. 10.8 which shows results involving estimation of the aerodynamic derivatives L_v, L_p and N_r for a SA-330 Puma helicopter for forward speeds of 60 and 100 knots. For the parameters L_v and N_r the trends in the theoretical and estimated values show a fairly close match, allowing for experimental errors. Results for the parameter L_p show similarity in terms of the trend but there is a large difference in terms of absolute values which might require further investigation and, possibly, changes within the nonlinear model.

The polar type of diagram mentioned in Chap. 3 allows a number of different measures to be compared for different models or comparisons to be made between

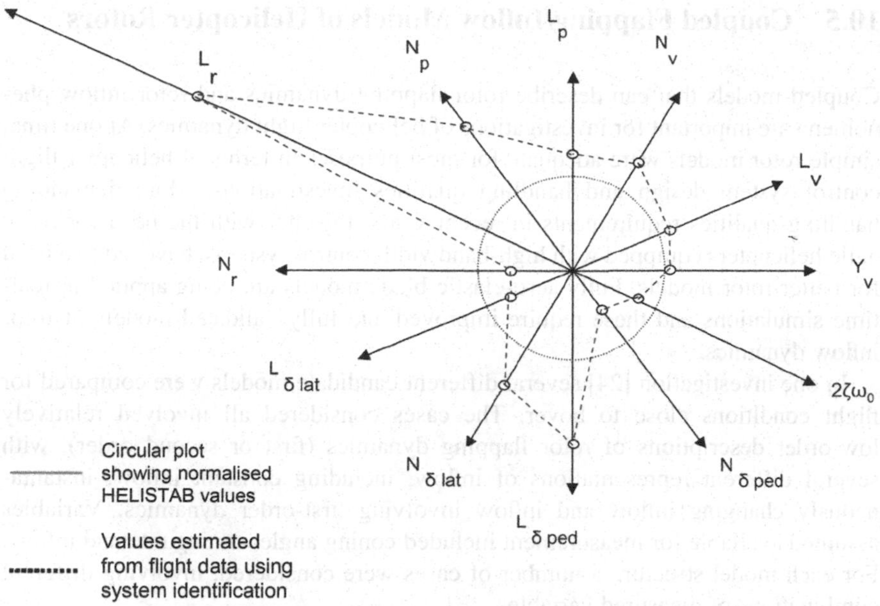

Fig. 10.9 Estimated and theoretical quantities for a model for the lateral/directional characteristic of a SA-330 Puma helicopter. The values obtained from the theoretical flight mechanics model (HELISTAB) [23] have been normalised to unity in each case and are shown by the *circle* in the *middle* of the polygon. Corresponding values estimated using system identification and parameter estimation methods are shown as individual points on the *radial lines* and these points are joined by the *dashed lines* to form a polygon. The parameters, measures and notation shown in this diagram are consistent with conventions normally adopted in helicopter flight mechanics modelling

estimated parameters from flight data and equivalent "theoretical" parameters, most often obtained by linearisation of the underlying nonlinear model for an equivalent flight condition. Figure 10.9 shows an example of the use of a polar diagram of this kind for such a comparison. In this case the theoretical values are normalised to unity and the estimated values from flight data are a multiple of that. It can be seen that one of the parameters estimated from flight is about four times the value predicted by linearisation of the flight mechanics model and this clearly raises important issues about the model that would justify further investigation. Estimated and theoretical values shown in this diagram are from data published in the AGARD Advisory Report 280 [7].

This example shows how the polar type of diagram can assist in the visualisation of issues that can arise with complex multi-parameter situations. Here the significance of the differences between the estimated and theoretical values of some specific parameters such as L_r,N_p and L_δ is clearly demonstrated. Further consideration of possible underlying reasons for these apparent errors is clearly needed.

10.5 Coupled Flapping/Inflow Models of Helicopter Rotors

Coupled models that can describe rotor flapping dynamics and rotor inflow phenomena are important for investigations of helicopter flight dynamics. At one time, simple rotor models were adequate for most purposes in terms of helicopter flight control system design and handling qualities investigations. More demanding handling qualities requirements in recent years, together with the need for more agile helicopters equipped with high-bandwidth control systems, have led to a need for better rotor models. Fully aero-elastic blade models are being applied in real-time simulations and these require improved and fully validated models of rotor inflow dynamics.

In one investigation [24] several different candidate models were compared for flight conditions close to hover. The cases considered all involved relatively low-order descriptions of rotor flapping dynamics (first or second order), with several different representations of inflow, including constant inflow, instantaneously changing inflow and inflow involving first-order dynamics. Variables assumed available for measurement included coning angle, coning rate and inflow. For each model structure a number of cases were considered, involving different combinations of measured variables.

Analysis of structural identifiability was completed for each of the configurations considered and gave useful information about the significance of the measured quantities for parameter estimation and model validation. For example, in some cases the model may be structurally unidentifiable in the absence of inflow measurements.

Optimal test inputs were found for all the cases considered through the use of simulation experiments involving the addition of measurement noise. Results showed that, in the absence of inflow measurements, some of the more complex models do not provide any more information than a very simple first order model having no inflow dynamics [24]. This result may explain why difficulties have been encountered in attempting to apply experimental modelling methods in the development of coupled flapping/inflow models [24]. This application provides an interesting illustration of the fact that the available output measurements for a parameter estimation or model validation study may be at least as important as the assumed model structure when determining the best test input.

10.6 Discussion

Particular problems in the development and testing of simulation models of helicopters include the fact that these vehicles operate in a non-uniform flow field and can show markedly nonlinear characteristics. Available experimental test records may also be short due to marginally stable or unstable dynamics under open-loop test conditions and records may also show high levels of sensor noise due to the high-vibration environment. Understanding model deficiencies and limitations requires us to relate parameters of identified models to more fundamental quantities within the vehicle model, such as moments of inertia and aerodynamic parameters. Implicit relationships between parameters of linearised models must also be investigated and understood if useful physical insight is to be gained.

Although system identification techniques undoubtedly have value for helicopter flight testing, for parameter estimation and for model validation purposes, problems remain and the full benefits of these techniques have not yet been realised. Difficulties relate particularly to issues of model "robustness" including [25]:

- robustness and reliability of *a priori* information used in the system identification process,
- robustness of the model structure, as determined from *a priori* information and from the results of identification,
- robustness of estimates of model parameters within the model structure determined above,
- robustness of the complete model.

Although the values of the variances associated with parameter estimates are useful indicators of robustness, it is important to understand that any comparison of variances obtained for different model structures is impossible. Checks of residuals can be important in that they can indicate deficiencies in terms of the chosen model structure (as discussed in Chap. 7) and an interesting indication of the robustness of parameter estimates may be obtained by repeating the estimation process using a number of records from the same flight experiment involving different record lengths. Plots for each parameter value estimated from these records versus the length of the experimental record used in the identification process can reveal useful information. Estimated parameter values should, in each case, tend towards a more or less constant value as the record length is increased and failure to do so shows poor robustness of estimates which may be related to problems of experiment design. Figure 10.10 shows an example of the behaviour of the estimates for one particular parameter within a six-degrees-of-freedom linearised model of a SA-330 Puma helicopter as the record length is varied. This is for one specific flight experiment, which, in this case involved a doublet input. The estimated value of the parameter M_q is seen to be positive for short experimental records, but with large values of the variance on the estimates. As the length of the experimental record is increased the estimated values become smaller and eventually tend towards a value of about -0.25, with very much smaller values of variance.

Understanding how parameter estimates change as the frequency range of measured variables is changed may also provide useful insight in terms of model robustness. High sensitivity of estimates to frequency over a part of the relevant range of frequencies may point to structural problems within the model or to inadequate experiment design. In general terms, maximising the frequency range over which estimates are more or less constant is clearly desirable.

The use of inverse simulation models (as discussed in Sect. 1.5.3 and also in Sect. 9.5.4 in the case study relating to the coupled-tanks system) has been

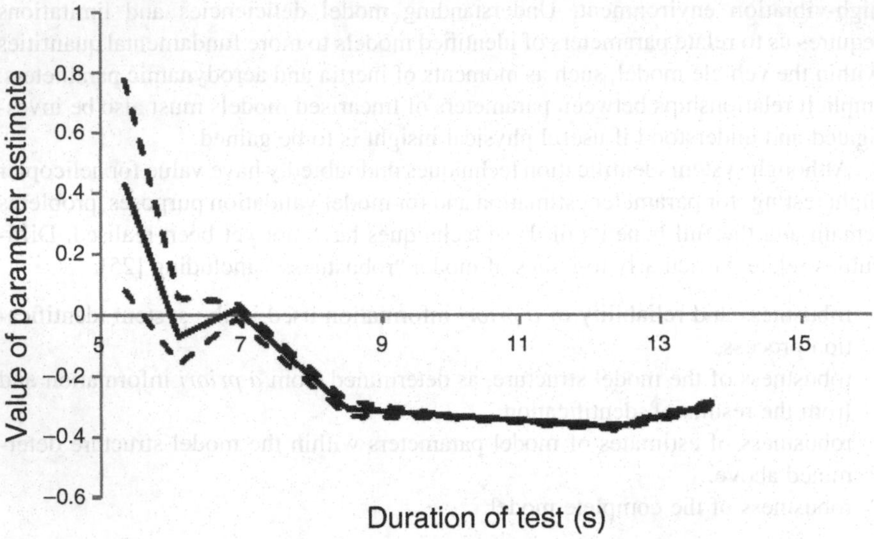

Fig. 10.10 Variation of estimates for parameter *Mq* with the length of the experimental record (*continuous lines*) for a SA-330 Puma helicopter. The magnitude of the corresponding standard errors of the estimates is indicated by the *dashed lines*. The model structure in this case was a linearised six-degrees of freedom model and the flight experiment involved an initial condition corresponding to straight and level flight with a forward speed of 100 knots. Identification was carried out in the frequency-domain and the upper limit of frequency used in the identification was 0.56 Hz. Measured data had been subjected to pre-processing to check compatibility of the signals from different sensors, to correct subsequently for measurement errors and to evaluate the data quality (Figure created from data published in [16])

suggested for helicopter flight mechanics model applications, especially in the context of system identification and model validation (see, e.g. [12]). However, published results do not appear to be available from any investigations of that kind and this is a potentially interesting area for research.

References

1. Hamel PG (1994) Aerospace vehicle modelling requirements for high bandwidth flight control. In: Cook MV, Rycroft MJ (eds) Aerospace vehicle dynamics and control. Clarendon, Oxford, pp 1–31
2. Anonymous (1994) Aeronautical design standard ADS-33D. Handling qualities specifications for military rotorcraft. Directorate for Engineering, US Army Aviation and Troop Command, St. Louis
3. Manness MA, Murray-Smith DJ (1992) Aspects of multivariable flight control law design for helicopters using eigenstructure assignment. J Am Helicopter Soc 37(3):18–32
4. Padfield GP, Du Val RW (1991) Application areas for rotorcraft system identification: simulation model validation. In: AGARD lecture series 178, Rotorcraft system identification, 12.1–12.30, AGARD, Neuilly-sur-Seine

5. Tischler MB, Remple RK (2006) Aircraft and rotorcraft system identification, Chapter 14 (Time-domain verification of identified models). AIAA, Reston, VA
6. Lu L, Padfield GD, White M et al (2011) Fidelity enhancement of a rotorcraft simulation model through system identification. Aeronaut J 115(1170):453–470
7. Anonymous (1991) Rotorcraft system identification, AGARD Advisory Report 280 (AGARD-AR-280). AGARD, Neuilly sur Seine
8. Rabiner LR, Schafer R, Rader CM (1969) The chirp z-transform algorithm. IEEE Trans Acoust Electroacoust AU17:86–92
9. Tischler MB, Remple RK (2006) Aircraft and rotorcraft system identification. AIAA, Reston, VA
10. Padfield GD, Thorne R, Murray-Smith DJ et al (1987) U.K. research into system identification for helicopter flight mechanics. Vertica 11(4):665–684
11. Anonymous (1991) Rotorcraft system identification, AGARD lecture series 178 (AGARD-LS-178). AGARD, Neuilly sur Seine
12. Bradley R, Padfield GD, Murray-Smith DJ et al (1990) Validation of helicopter mathematical models. Trans Inst Meas Control 12(4):186–196
13. Hunter WG, Hill WJ, Henson HL (1969) Designing experiments for precise estimation of some of the constants in a mechanistic model. Can J Chem Eng 47:76–80
14. Maine R, Iliffe R (1985) Identification of dynamic systems – theory and implementation. NASA report RP-1138. NASA, Dryden Flight Research Center, Edwards, CA
15. Klein V (1979) Identification evaluation methods. In: AGARD lecture series 104 (AGARD-LS-104) Rotorcraft system identification, Section 2. AGARD, Neuilly-sur-Seine
16. Black CG, Murray-Smith DJ (1989) A frequency-domain system identification approach to helicopter flight mechanics model validation. Vertica 13(3):343–368
17. Rao CR (1945) Information and the accuracy attainable in the estimation of statistical parameters. Bull Calc Math Soc 37:81–91
18. Tischler MB (1991) Identification techniques –frequency domain methods. In: AGARD lecture series 178 (AGARD-LS-178) Rotorcraft system identification, Section 6. AGARD, Neuilly-sur-Seine
19. Kaletka J (1979) Rotorcraft identification experience. In: AGARD lecture series 104 (AGARD-LS-104) Rotorcraft system identification, Section 7. AGARD, Neuilly-sur-Seine
20. Otnes RK, Enochson J (1978) Applied time series analysis. Wiley, New York
21. Leith DJ, Murray-Smith DJ (1989) Experience with multi-step test inputs for helicopter parameter identification. Vertica 3(3):403–412
22. Mehra RK (1974) Optimal input signals for parameter estimation in dynamic systems – survey and new results. IEEE Trans Auto Control AC-19:753–768
23. Padfield GD (1981) A theoretical model of helicopter flight mechanics for application to piloted simulation. Royal Aircraft Establishment Technical Report 80148. HMSO, London
24. Bradley R, Black CG, Murray-Smith DJ (1989) System identification strategics for models incorporating induced flow. Vertica 13(3):281–293
25. Murray-Smith DJ (1991) Robustness issues. In: AGARD Advisory Report 280, Rotorcraft system identification, (AGARD-AR-280). AGARD. Neuilly-sur-Seine, pp 213–222

Chapter 11
Case Study: Compartmental Modelling of the Gas-Exchange Processes of the Human Lungs

11.1 Introduction

The respiratory and the cardiovascular systems together ensure that the brain and other body tissues receive an adequate supply of oxygen. At the same time, these systems also remove the carbon dioxide resulting from body activity so that acidity levels within the system are at appropriate levels. The processes of gas exchange between blood and air within the lungs and also between the blood and body tissues are of central importance for this. Within the lungs ventilation is cyclic and involves two main phases. The first of these is inspiration, during which fresh air is delivered to the alveoli of the lungs where the oxygen can diffuse into the blood through a thin membrane. The second phase involves expiration when carbon dioxide diffuses into the alveolar gas mixture from the blood. The alveoli of the lungs have a large surface area and a very thin membrane forms the blood-gas barrier. The partial pressure of oxygen is greater in alveolar air than in venous blood and, conversely, the partial pressure of carbon dioxide in venous blood is higher than in alveolar air, resulting in the exchange of these gases.

While it is clear that gas exchange is central to the function of the respiratory system, the main function of the cardiovascular system is the transport of gases between the lungs and all the other organs of the body. This process involves a protein within the red blood cells known as haemoglobin. Oxygen exists in the blood in two distinct forms. One of these, which is a small component, is oxygen dissolved in plasma while the other involves oxygen which has combined with haemoglobin to form oxyhaemoglobin. Carbon dioxide is also transported by the blood in two forms. The first of these is carbon dioxide dissolved in the plasma and the other is in a chemically combined form through the action of an enzyme known as carbonic anhydrase.

Control processes within the respiratory and cardiovascular systems are complex and involve a number of controlled variables (such as arterial carbon dioxide and

D.J. Murray-Smith, *Testing and Validation of Computer Simulation Models*,
Simulation Foundations, Methods and Applications,
DOI 10.1007/978-3-319-15099-4_11

arterial oxygen) and several other variables through which control is achieved (including the breath size and the frequency of breathing). Components within the peripheral and central nervous systems are also of key importance, such as the peripheral chemo-receptors in the arteries and the central chemo-receptors which are located within the brain close to the respiratory centre which is believed to have an important coordinating role.

Continuous measurement of variables within the system in human subjects is limited significantly by technical and ethical factors. Standard measurements that can be made of variables within the human respiratory system in a non-invasive fashion include air flow at the mouth, changes in body volume using body-plethysmography techniques, and analysis of the expired gas mixture to provide measures of partial pressures of the individual component gases.

A useful outline of respiratory physiology, prepared as part of an integrated systemic view of the use of mathematical models and computer simulation techniques through which respiratory measurements can be interpreted, has been provided by Petersen and Cunningham [1]. Within the edited volume containing that review are other chapters on clinical measurements [2], mathematical studies of respiratory phenomena [3], modelling processes in respiratory medicine [4] and case studies of respiratory system models [5]. Although published in 1988 these still provide much useful material for anyone wishing to explore this field in more detail.

11.1.1 Modelling and Simulation Techniques in Respiratory Physiology

Dynamic models describing the gas exchange processes of the lungs are of considerable interest and can provide insight at many levels, from cellular to organ and whole-body situations. Models can therefore be very useful as an aid in the teaching of respiratory physiology as well as in research. Diseased states can be investigated through modelling and simulation techniques and validated models can even provide potentially important tools for clinical decision making. Gas exchange models also provide a basis for the development of improved procedures for clinical tests of pulmonary function.

A review paper by Ben-Tal (2006) [6] presents a hierarchy of models of increasing complexity for gas exchange processes within the human lungs. This hierarchy allows links to be established between models describing processes at the molecular level and "systems-level" descriptions which are based on sub-models describing the action of individual organs such as the lungs, heart and tissues. Another valuable account of gas exchange modelling issues is provided by the paper of Hahn and Farmery (2003) [7].

One important feature is that, for clinical applications, models must have a structure and parameters that are appropriate for the human subject under investigation. This means that although a general form of gas exchange model may be established initially using physical and physiological principles, system identification and parameter estimation methods are essential for the development of models for individual subjects. A number of studies have also been reported which have been concerned with the use of dynamic models and system identification methods for the development of non-invasive techniques for the estimation of physiological quantities and this topic is considered further in Sect. 11.3.

Both the structure and the parameter values of the model must be appropriate and, if it is fit for purpose, the model should have a set of outputs that match closely a corresponding set of measured variables. Structural variations within pulmonary gas exchange models can often be associated with disorders or abnormalities such as circulatory shunts. The choice of model structure may be at least as important, in terms of the intended application of a model, as any parametric issues. Both the need to model individual subjects and the biological variability that is found in physiological models are factors in this case study that are typical of many other dynamic modelling situations involving biomedical applications.

11.2 Simple Compartmental Models with Continuous Gas and Blood Flow

Compartmental models are very widely used in the modelling of physiological systems. They provide a physically and anatomically based foundation for the development of mathematical descriptions that lead to simple and intuitive interpretations. Compartmental models are also convenient in a mathematical sense since they often result in models that have a state-space structure that allows powerful analytical tools to be applied. However, care has to be taken in the development of compartmental models that mathematical convenience does not lead to over-simplification.

The gas exchange properties of the human cardio-respiratory system provide an interesting illustration of how different assumptions and simplifications in terms of the model structure can lead to very different forms of compartmental model and, potentially, to difficulties unless the models are used with great care. The case considered in this section relates to compartmental models involving continuous flow of gases and of blood and neglects the tidal nature of gas flow and the fact that blood flow is also not truly continuous but involves a rhythmic component driven by the heart. This is clearly a highly simplified representation of the anatomy and physiology of the lungs. An alternative compartmental representation involving tidal breathing is described in Sect. 11.3.

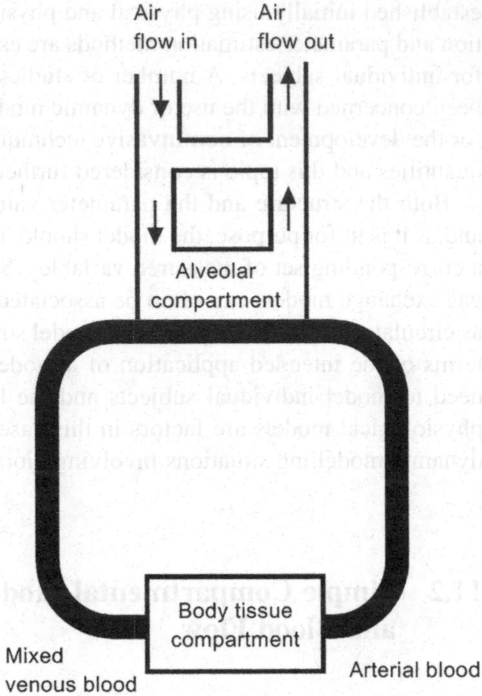

Fig. 11.1 Schematic diagram of continuous gas-flow and continuous blood-flow type of gas exchange model

11.2.1 General Case for Gases That Are Soluble in Blood

The conventional continuous gas flow and continuous blood flow type of model which has been used for the representation of gas exchange for a variety of gases, including inert gases is shown in diagrammatic form in Fig. 11.1. This type of representation is one of the simplest descriptions considered in the papers by Ben-Tal [6] and Hahn and Farmery [7]. The diagram of Fig. 11.1 shows gas and blood flow for a lung model involving a single alveolar compartment with volume $V_A$and alveolar ventilation $\dot{V}_A$ and a parallel dead-space pathway having a gas flow rate $\dot{V}_D$. The blood flow associated with the alveolar compartment is $\dot{Q}_P$ and the structure of the model can accommodate a parallel blood shunt pathway, although this is not shown in Fig. 11.1.

The equation relating the rate of change of mass of the chosen gas in the alveolar compartment to the input and output mass flows is:

$$V_A \frac{dF_A}{dt} = \dot{V}_A(F_I - F_A) + \dot{Q}_P(C_a - C_{\overline{V}}) \tag{11.1}$$

where F represents the fractional concentration of the gas being considered, C represents the blood gas content, $\dot{Q}_P$ is the blood flow associated with the alveolar compartment and $\dot{V}_A$ is the corresponding gas flow. The quantity F_I is the fractional concentration of the chosen gas in the inspired mixture and F_A is the fractional concentration of the gas in the alveolar compartment. The variable C_a is the end-capillary blood gas content and $C_{\overline{V}}$ is the mixed-venous blood gas content. All the variables, F_I, F_A, C_a and $C_{\overline{V}}$ are, in general, time varying quantities.

The first term on the right hand side of (11.1) represents the net change of mass of the chosen gas in the alveolar compartment, while the second term represents the net flow of gas across the alveolar-capillary membrane (the blood-gas barrier). Commonly used input variables for this model are F_I and $\dot{V}_A$. The input most suitable for manipulation to mimic experimental conditions is the inspired gas concentration F_I and this can be switched from one level to another in a step-wise fashion to represent the switching of the subject from one gas mixture to another. This input variable can also be changed in a sinusoidal or other more complex fashion to mimic the operation of a controlled gas mixing system. Manipulation of the gas concentration input can also be valuable for the purposes of system identification and parameter estimation.

It should be noted that in Fig. 11.1 the dead-space is represented by a gas flow pathway which lies in parallel with the alveolar compartment. This is, of course, anatomically incorrect since no gas mixing takes place in this dead space and it is a significant simplification. A relationship between the parallel dead space flow ($\dot{V}_D$) and the total gas flow ($\dot{V}_T$) can be found by looking at condition that gives a balance in terms of gas flow rates and alveolar, inspiratory and mean expiratory gas concentrations. This gives an equation:

$$\frac{\dot{V}_D}{\dot{V}_T} = \frac{F_{\overline{E}} - F_A}{F_I - F_A} \tag{11.2}$$

The Eq. (11.2) is a widely used relationship, often involving carbon dioxide as the chosen gas, which has been applied in clinical tests of respiratory function for some considerable time (see, e.g. [7]). It is important to note that the derivation of this equation involves a balanced condition for gas flows rather than for gas volumes and it is conventionally assumed that these quantities are related in a simple linear way so that the ratio of gas flows on the left hand side of the equation has the same magnitude as the corresponding ratio of gas volumes.

The model may be modified to include a blood shunt pathway where a part of the total blood flow bypasses the alveolar compartment. This feature is not shown in Fig. 11.1, but a shunt pathway of this kind is included in the model shown in Fig. 11.8 and would be the same in this case. The equation for this is closely analogous to (11.2) and has the form:

$$\frac{\dot{Q}_S}{\dot{Q}_T} = \frac{C_a - C_{\overline{a}}}{C_a - C_{\overline{V}}} \tag{11.3}$$

where $\dot{Q}_s$ is the blood flow in the shunt pathway, $\dot{Q}_T$ is the total blood flow in the shunt pathway C_a is the end-capillary blood gas content, $C_{\bar{a}}$ is the mixed-arterial blood gas content and $C_{\overline{V}}$ is the mixed-venous blood gas content. This equation is conventionally used in clinical applications with oxygen as the marker gas. It is therefore nonlinear in terms of partial pressure measures because the oxyhaemoglobin equilibrium relationship has a nonlinear form.

11.2.2 Case of Insoluble Inert Gases

In the case of an inert gas which has negligible solubility in blood the second of the two terms on the right-hand side of (11.1) disappears and the model is described by the equation:

$$V_A \frac{dF_A}{dt} = \dot{V}_A(F_I - F_A) \tag{11.4}$$

In some variations of this model the terms involving fractional concentrations, denoted by F above, are replaced by equivalent terms involving partial pressures, P, since these types of measure are linearly related for inert gases. The form of model given in (11.4) is very important in clinical applications and forms the basis of a number of tests of lung function. Many widely-used models of anaesthetic gas uptake and elimination processes are also based on simple lumped-parameter models of this kind.

11.2.3 Issues of Model Quality in Models with Continuous Ventilation and Perfusion

The attraction of the continuous ventilation and perfusion models of the kind discussed above in Sects. 11.1.1 and 11.1.2 is that they are essentially linear in form and thus convenient in terms of mathematical solutions. They describe processes in a way that allows simple analogies to be drawn with physical systems. They have also been used extensively in teaching and with much emphasis on analysis of steady-state conditions. Unfortunately, they also tend to be misleading in that the model structure is unlike any possible simplified structure for the human lungs. As pointed out by Hahn and Farmery [7], the structure for these models is much closer to that of the fish gill where water flows continuously over filaments which have laminae through which the blood flows in a continuous fashion. In the continuous ventilation and perfusion models of the human lung the water flow is replaced by gas flow but the realities of the gas exchange processes of the human lungs are very different from those of the water-blood interface of the gill. The processes of the human lung are cyclic in nature, with the period of the complete breath cycle divided into inspiratory and expiratory periods. Air breathed in at the mouth passes through upper airways before reaching the alveoli where gas exchange with the blood can take place. This means that there should be a dead-space compartment with tidal ventilation, rather than the dead-space flow of the continuous ventilation model.

The introduction of tidal ventilation into gas exchange models, with separate inspiratory and expiratory periods, inevitably means that the mathematical form of the models becomes more complex and that linear methods of analysis can no longer be applied. However, the potential value of moving to the anatomically more realistic tidal ventilation type of model is considerable. This form of model is more compatible with observations of the real system and the availability of easily applied simulation tools makes the task of creating a computer-based model with a convenient user interface a relatively trivial one. This means that, for a wide range of applications in physiological and medical education and in the development of clinical tests in respiratory medicine, the restrictions of linear models and steady-state analysis can be forgotten.

The paper by Hahn and Farmery [7] is a very interesting and hard-hitting review of gas exchange models. The conclusion presented in that paper was that the use of continuous ventilation models for the human lungs should be abandoned, together with the use of unjustified steady-state assumptions. Instead, more effort should be placed on improving nonlinear tidal-ventilation models.

11.3 A Compartmental Model with Tidal Breathing

Representations of the normal lungs that involve tidal ventilation rather than the continuous ventilation type of description of Sect. 11.2 above are clearly much more reasonable in terms of the known anatomy and physiology of the human respiratory system. A typical model of this kind involves a lumped-parameter description with a single uniform and compliant alveolar compartment, a single compartment which represents the body tissues and a rigid non-reacting dead-space compartment which represents the conducting airways. The tissue and alveolar compartments are linked by the blood circulation and the model incorporates the cyclic ventilation processes associated with breathing. The structure of this highly simplified compartmental model is shown in Fig. 11.2 (see, e.g. [5, 8–10]).

Symbols used in describing the model of Figs. 11.1 and 11.2 in quantitative form are consistent with conventions commonly used in respiratory physiology [11]. It should also be noted that, in the model incorporating tidal ventilation, the breath cycle is divided into inspiratory and expiratory periods of duration T_I seconds and T_E seconds respectively.

Ordinary differential equations for this model structure may be derived using the principle of conservation of mass. In the case of carbon dioxide, the exchange of gas between the blood and the alveolar compartment during inspiration is given by the equation:

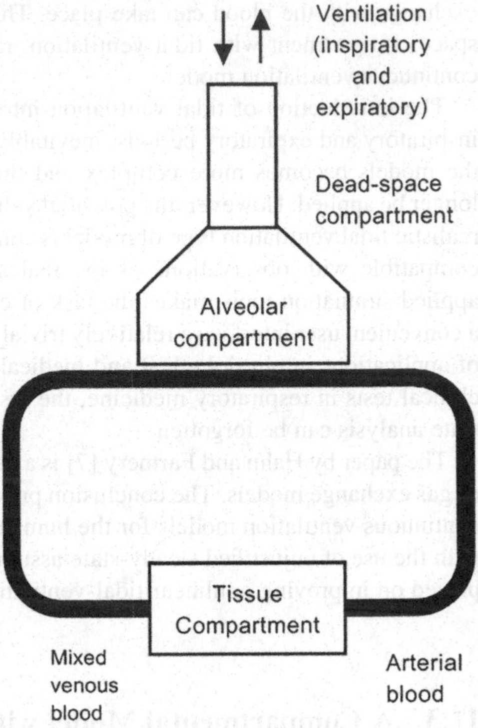

Fig. 11.2 Schematic diagram of gas exchange model involving a single homogeneous alveolar compartment and a dead-space compartment with tidal ventilation. Blood flow is assumed to be continuous

$$\frac{d(F_A V_A)}{dt} = \dot{V} F_{IA} + \dot{Q}_P [C_{\overline{v}} - C_{\overline{a}}] \tag{11.5}$$

Here the variable F_{IA} is a function of time representing the concentration of the gas of interest in the inspired mixture. The variable $\dot{V}$ is the flow rate during inspiration (i.e. for $\dot{V} \geq 0$) and corresponds to the time rate of change of V_A, with a fixed end-expiratory volume. There is a difficulty caused by the fact that this variable is measured at the mouth and the gas mixture reaching the alveolar compartment at the start of the inspiratory phase always involves gas from the dead-space. This dead-space gas has come from the alveolar compartment in the final stage of expiration during the previous breath cycle. There is, thus, always an element of re-breathing from the dead-space compartment at the start of each breath cycle. The variable $F_{IA}(t)$ therefore varies during the inspiratory part of the breath cycle as the inspired gas passes through the dead space volume. The alveolar volume changes with time and may be related to the measured gas flow at the mouth or there may be an assumption about the form of flow pattern during inspiration.

During expiration $\left(\dot{V} < 0\right)$ the mass-balance condition is given by:

$$\frac{d(F_A V_A)}{dt} = \dot{V} F_A + \dot{Q}_P[C_{\bar{v}} - C_{\bar{a}}] \tag{11.6}$$

and since $\dot{V} = \frac{dV_A}{dt}$ this equation may be reduced to:

$$V_A \frac{\mathrm{d}FA}{\mathrm{d}t} = \dot{Q}_P[C_{\bar{v}} - C_{\bar{a}}] \tag{11.7}$$

The variation in the volume of the alveolar compartment may again be found by integration of the measured flow $\dot{V}(t)$ with respect to time or may be assumed to have a specific form as a function of time during this phase of the breath cycle.

The terms on the on the left hand sides of (11.5) and (11.6) represent the rate of change of mass of gas within the alveolar compartment, while the terms on the right hand side represent, in each case, the rate of mass transfer to the alveolar compartment of gas from the dead space and the blood.

Similarly, for the tissue compartment representing the rest of the body, the rate of change of mass of the chosen gas within the tissue compartment must be balanced by the difference between the consumption or metabolic production of the gas and the rate of mass transfer through the circulation to the alveolar compartment. This gives a third equation:

$$V_{TC} \frac{dC_{\bar{v}}}{dt} = \dot{Q}_{TC}[C_{\bar{v}} - C_{\bar{a}}] - \psi \tag{11.8}$$

where $\psi(t)$ represents the "consumption" of the gas under consideration. If ψ is negative it represents a metabolic production of gas within the tissues, as applies in the case of carbon dioxide.

Since the alveolar compartment is considered to be compliant, the alveolar compartment volume V_A is not constant and varies throughout the breath cycle. This volume is given by the equation:

$$V_A(t) = V_A(0) + \Delta V_A(t) \tag{11.9}$$

where $V_A(0)$ is the initial value of the alveolar compartment volume and $\Delta V_A(t)$ is the change of volume due to the flow $\dot{V}(t)$ and can be found by integrating $\dot{V}(t)$ with respect to time.

It should be noted that the equations for this model, as with the continuous-ventilation representation of Sect. 11.1, involve the quantities C_a and $C_{\bar{v}}$ which are quantities based on blood gas content. Relating these variables to measurements in terms of partial pressures involves a conversion process that depends on the gas in question. In the case of oxygen this is relatively complex since it involves the nonlinear oxyhaemoglobin equilibrium curve. In the case of carbon dioxide the relationship is somewhat simpler but can be nonlinear over parts of the operating range. Other gases may involve a linear relationship or may be almost insoluble in blood which, as seen in the derivation of (11.4), can result in a significant simplification.

It is of interest to note that, although the tidal-ventilation form of model is very different in physical structure compared with the structure of the continuous ventilation model, it is possible to derive an expression for the ratio of dead-space volume to tidal volume which is remarkably similar to (11.2) which applies to the continuous form. In the tidal-ventilation case the form of the expression is:

$$\frac{V_D}{V_T} = \frac{F_{\overline{E}} - F_{\overline{A}}}{F_I - F_{\overline{A}}} \tag{11.10}$$

It should be noted that, in this case, the left-hand side of the equation involves the ratio of volumes, while in (11.2) the relationship involves a ratio of flows. Note also that the variable $F_{\bar{A}}$ in (11.10) is the mean concentration in the alveolar compartment over the period of the expiration, T_E.

The equation for the blood shunt pathway, if one exists, is exactly the same in the tidal model as in the continuous version and has the form:

$$\frac{\dot{Q}_S}{\dot{Q}_T} = \frac{C_a - C_{\overline{a}}}{C_a - C_{\overline{v}}} \tag{11.11}$$

The only difference from the case involving continuous ventilation is that averaging of the time-varying blood gas contents variables must now take place over the complete period of the breath cycle ($T_I + T_E$).

11.3.1 A Tidal-Ventilation Model for Carbon Dioxide

In the case of carbon dioxide, with gas concentration expressed in terms of partial pressure, the model has to incorporate an appropriate representation for the dissociation curve for carbon dioxide in blood. For the range of partial pressure values of interest in the alveolar and tissue compartments, the dissociation curves for mixed venous and arterial blood may be represented by parallel straight lines. The gradient parameter b (mmol/l) for these straight line approximations is known to be related to the haemoglobin concentration Hb (g/100 ml) by the following equation:

$$b \cong 0.703 + 0.050\,Hb \tag{11.12}$$

The dissociation curve representation involves a second parameter a_d which is the difference in intercepts for the individual dissociation curves for mixed venous and arterial blood. It is known that the value of this parameter varies slightly with the level of oxygen in the blood and the value chosen for this quantity within the gas exchange model is 0.58 mmol/l, which is believed to be an appropriate value for the conditions considered [12].

Use of a linear approximation for the carbon dioxide dissociation curves leads to a simplification of (11.5), (11.6), (11.7), and (11.8) to give:

$$\frac{dP_A}{dt} = \frac{S\dot{V}}{V_A}\left(P_I^* - P_A\right) + \frac{\dot{Q}}{V_A}\left[a_d + b(P_{TC} - P_A)\right]\gamma \tag{11.13}$$

$$\frac{dP_{TC}}{dt} = \frac{\dot{M}}{bV_{TC}} - \frac{\dot{Q}}{bV_{TC}}\left[a_d + b(P_{TC} - P_A)\right] \tag{11.14}$$

where γ is a constant of known value and the quantity $P_I^*(t)$ has a form that varies within the breath cycle. This can be divided into three distinct stages. Stage 1 involves transfer to the alveolar compartment of gas which was in the dead space at the end of the previous breath cycle. The condition defining this stage is:

$$\dot{V}(t) \geq 0 \quad \text{and} \quad \int_{t_I}^{t} \dot{V}(t)dt \leq V_D \tag{11.15}$$

where t_I defines the start time of the inspiratory phase.

Stage 2 involves transfer of gas inspired during the current breath cycle to the alveolar compartment from the dead space and the corresponding condition is:

$$\dot{V}(t) \geq 0 \quad \text{and} \quad \int_{t_I}^{t} \dot{V}(t)dt > V_D \tag{11.16}$$

In Stage 3 expiration takes place, for which the condition is:

$$\dot{V}(t) < 0 \tag{11.17}$$

During Stage 1, $P_I^*(t) = P_D(t)$ where $P_D(t)$ is the partial pressure of gas entering the alveolar compartment and is evaluated as a flow-weighted mean of the alveolar partial pressure $P_A(t)$ over the final (end-tidal) portion of the previous expiration. During the remainder of the inspiratory phase of the breath cycle (Stage 2), $P_I^*(t) = P_I(t)$, where $P_I(t)$ is the partial pressure of carbon dioxide in the inspired gas mixture.

Descriptions based on this model structure have been shown, in a study by Bache and his colleagues [13], to be capable of describing gas-exchanging processes of the normal human lungs for experiments of duration about 10 min or less, for carbon dioxide, oxygen and other gases. More information about the assumptions made in using a lumped parameter compartmental model of this kind to represent the gas exchanging properties of the lungs and the approximations applied in the derivation of these equations may be found elsewhere (see, e.g., [5, 10]).

Variables of the model that are of special interest are the ventilation $(\dot{V}(t))$ and the partial pressure of carbon dioxide in the inspired mixture ($P_I(t)$) and these two quantities, when taken together, provide a means of applying test inputs to the system. The partial pressure of carbon dioxide in the alveolar compartment ($P_A(t)$) is also of particular interest and may be regarded as an output variable. This is difficult to measure in any direct fashion but the measured partial pressure of carbon dioxide at the mouth, taken over the final ("end-tidal") phase of each expiration, can be used to provide sampled values which approximate to the alveolar partial pressure for the final phase of each breath cycle, provided account is taken of the inevitable transport delay as the expired gases flow through the upper airways (represented in the model by the dead space compartment). The variable $P_{TC}(t)$ in (11.14) is equivalent to the partial pressure of mixed venous blood and can be regarded as a second output variable, but this cannot be measured so readily.

11.3.2 Identifiability Analysis of the Gas Exchange Model for Carbon Dioxide

As discussed in Chap. 5, issues of global and pathological unidentifiability are important both for the successful use of experimental modelling techniques, such as system identification and parameter estimation and for model validation. Experimental design and the selection of test inputs for practical system testing depend on concepts closely linked to identifiability. Through the use of the techniques presented in Chap. 5, investigation of the global identifiability may be carried out using a linearised version of the nonlinear model for carbon dioxide based on tidal ventilation given in Sect. 11.3.1.

Applying the Laplace transformation and carrying out simple algebraic manipulations the model may be shown to take the form:

$$P_A(s) = \frac{b_1 s + b_2}{s^2 + a_1 s + a_2} U(s) + \frac{c_1 s + c_2}{s^2 + a_1 s + a_2} + \frac{d_1}{s} \tag{11.18}$$

where $P_A(s)$ is the Laplace transform of the variable $P_A(t)$ and the variable $U(s)$ is the Laplace transform of an input variable that may be defined in terms of the effective ventilation variable $\dot{V}_E(t)$ and the partial pressure of carbon dioxide $P_I(t)$ in the inspired gas mixture. The coefficients appearing within the numerator and denominator of each term on the right hand side of (11.18) may be shown to be related to the parameters of (11.13) and (11.14) as follows:

$$a_1 = \frac{\dot{Q}}{V_{TC}} + \frac{\dot{V}_E}{V_A(0)} + \frac{k\dot{Q}b}{V_A(0)} \tag{11.19}$$

$$a_2 = \frac{\dot{V}_E \dot{Q}}{V_{TC} V_A(0)} \tag{11.20}$$

$$b_1 = \frac{1}{V_A(0)} \tag{11.21}$$

$$b_2 = \frac{\dot{Q}}{V_{TC} V_A(0)} \tag{11.22}$$

$$c_1 = P_A(0) + \frac{k\dot{M}}{\dot{V}_E} \tag{11.23}$$

$$c_2 = \frac{\dot{Q}}{V_{TC}} P_A(0) + \frac{k\dot{Q}(a_d + bP_{TC}(0))}{V_A(0)} - \frac{k\dot{M}}{\dot{V}_E}\left[\frac{\dot{Q}}{V_{TC}} + \frac{\dot{V}_E}{V_A(0)} + \frac{k\dot{Q}b}{V_A(0)}\right] \tag{11.24}$$

$$d_1 = \frac{k\dot{M}}{\dot{V}_E} \tag{11.25}$$

where k is a constant.

The number of coefficients in the transformed model of (11.18) defines the limit in terms of the number of quantities that can be estimated independently from tests carried out on the system. Equally, this limits the information available for model validation purposes. Before relating these results to the parameters of the simulation model of (11.13) and (11.14) the original parameters must be associated with the coefficients of the transformed model. The method of approach in this case is based on the fact that the effective ventilation $\dot{V}_E$ is a measured quantity so that the coefficients a_2 and b_2 in (11.20) and (11.22) are found to be linearly related. Manipulation of the remaining algebraic equations for the coefficients of the transformed model gives the following set of relationships involving parameters of the original model:

$$V_A(0) = \frac{1}{b_1} \tag{11.26}$$

$$P_A(0) = c_1 - d_1 \tag{11.27}$$

$$P_{TC}(0) = \frac{c_2 - \frac{b_2}{b_1}(c_1 - d_1) + a_1 d_1}{a_1 - \frac{b_2}{b_1} - b_1 \dot{V}_E} - \frac{a_d}{b} \tag{11.28}$$

$$\dot{M} = \frac{\dot{V}_E d_1}{k} \tag{11.29}$$

$$b\dot{Q} = \frac{1}{kb_1}\left(a_1 - \frac{b_2}{b_1} - \dot{V}_E b_1\right) \tag{11.30}$$

$$bV_{TC} = \frac{1}{kb_2}\left(a_1 - \frac{b_2}{b_1} - \dot{V}_E b_1\right) \tag{11.31}$$

From (11.26), (11.27), (11.28), (11.29), (11.30), (11.31) it can be seen that knowledge of the parameters a_d and b which are used to describe the dissociation curves for carbon dioxide in mixed venous and arterial blood is essential if the other parameters and initial values of the lung gas exchange model are to be estimated through system identification and parameter estimation methods. Thus, provided a_d and b are known the model is globally identifiable [13, 14].

While identifiability analysis of this kind can help establish theoretical limits in terms of the estimates that may be possible using an experimental modelling approach, there are also important issues of pathological unidentifiability that have to be considered. This depends on experiment design and especially on the form of test input applied. In particular, if an input is used which does not excite the system sufficiently it is likely that some parameter estimates obtained from measured response data involving $P_A(t)$ will be poorly determined, with large error bounds. Test signal design methods of the type described in Chap. 5 can help to ensure well-defined estimates of parameters and retrospective analysis may also be carried out through examination of the sensitivity matrix $\boldsymbol{X}$ and the closely associated parameter information matrix $\boldsymbol{M}$. Pathological unidentifiability is associated with linear dependence of the columns of the sensitivity matrix which, in the case of the lung model, has the following form for measured response data involving m observations:

$$\boldsymbol{X} = \begin{bmatrix} \frac{\partial P_{A_1}}{\partial V_A} & \frac{\partial P_{A_1}}{\partial V_{TC}} & \frac{\partial P_{A_1}}{\partial \dot{Q}} & \frac{\partial P_{A_1}}{\partial \dot{M}} & \frac{\partial P_{A_1}}{\partial P_A(0)} & \frac{\partial P_{A1}}{\partial P_{TC}(0)} \\ \frac{\partial P_{A_2}}{\partial V_A} & \frac{\partial P_{A_2}}{\partial V_{TC}} & \frac{\partial P_{A_2}}{\partial \dot{Q}} & \frac{\partial P_{A_2}}{\partial \dot{M}} & \frac{\partial P_{A_2}}{\partial P_A(0)} & \frac{\partial P_{A_2}}{\partial P_{TC}(0)} \\ \vdots & \vdots & \vdots & \vdots & \vdots & \vdots \\ \frac{\partial P_{A_m}}{\partial V_A} & \frac{\partial P_{A_m}}{\partial V_{TC}} & \frac{\partial P_{A_m}}{\partial \dot{Q}} & \frac{\partial P_{A_m}}{\partial \dot{M}} & \frac{\partial P_{A_m}}{\partial P_A(0)} & \frac{\partial P_{A_m}}{\partial P_{TC}(0)} \end{bmatrix} \qquad (11.32)$$

Pathological unidentifiability occurs if any two columns of this sensitivity matrix $\boldsymbol{X}$ are linearly dependent. Unidentifiability or near-unidentifiability is also reflected in the condition number of the parameter information matrix $\boldsymbol{M}$ which (as discussed in Chap. 5) is defined as:

$$\boldsymbol{M} = \boldsymbol{X}^T \boldsymbol{X} \qquad (11.33)$$

The condition number is the ratio of the largest eigenvalue of $\boldsymbol{M}$ to the smallest and a large condition number indicates that the parameter estimates will be poorly defined. As discussed in Chap. 5 an alternative measure is provided by the determinant of $\boldsymbol{M}$ and small values of this measure correspond to poorly defined estimates.

A further useful measure of model quality is based on the parameter correlation matrix $\boldsymbol{P}$ which (as outlined in Chap. 5) may be defined in terms of its elements as:

$$p_{ij} = \frac{m_{ij}^{-1}}{\sqrt{m_{ii}^{-1} m_{jj}^{-1}}} \tag{11.34}$$

where P_{ij} is the element of the matrix $\boldsymbol{P}$ in row i and column j and m_{ij}^{-1} is the element of $\boldsymbol{M}^{-1}$ in row i and column j. The diagonal terms of this matrix are unity and off-diagonal terms lie between -1 and 1. Problems of unidentifiability may arise if any off-diagonal elements within $\boldsymbol{P}$ are close to unity, with a value of 0.95 being regarded as a practical limit [13]. Small values of off-diagonal elements suggest that parameters of the model are decoupled.

Preliminary results for the gas exchange model were obtained from a simple experiment involving the subject breathing air for 40s, followed by a sudden switch to a gas mixture containing 7 % carbon dioxide for a further 80 s. This form of test input is, essentially, a step function and analysis of the results in terms of the parameter correlation matrix $\boldsymbol{P}$ showed clear signs of pathological unidentifiability with some very large off-diagonal elements associated with the parameters $\acute{M}$ and V_{TC} (0.9999), $\acute{M}$ and V_A (0.840), V_A and V_{TC} (0.841) and $\acute{Q}$ and P_{TC} (0) (-0.952). These large values of off-diagonal elements of $\boldsymbol{P}$ indicate clearly that this is not an ideal form of test input [14]. A more persistently exciting input is required and this must not only have appropriate frequency content but the magnitude of carbon dioxide and other gases in the inspired gas mixture and the duration of the test must take full account of known physiological constraints.

11.3.3 *Experimental Constraints for the Gas Exchange Model for Carbon Dioxide*

Obvious constraints exist if experimental measurements are limited to gas flow rates and carbon dioxide partial pressures at the subject's mouth. In the experiments in the study described in [13] gas flow rates were measured using separate pneumotachographs for inspiratory and expiratory phases of the breath cycle. A respiratory mass spectrometer was used to provide continuous measurements of gas concentrations at the mouth. There was a transport delay of 220 ms within the mass spectrometer and a corresponding delay had to be introduced for the flow measurements in order to ensure that all measurements were synchronised.

If physiological constraints are ignored it might appear to be appropriate to use a long test period with an input involving a test signal with a high concentration of carbon dioxide. However, limits have to be imposed on the concentration of carbon dioxide since only relatively low levels of this gas can be tolerated with comfort. Constraints on the input concentration and experiment duration also arise because it is known that the cardiac output, which corresponds to the parameter $\dot{Q}$ of the model, is affected by inspiration of carbon dioxide. The available evidence suggests, however, that levels of carbon dioxide of the order of 5 % in the inspired mixture have insignificant effects on the cardiac output if breathed for a few minutes only. Therefore, the gas mixture chosen for the investigation outlined here involved 5 % carbon dioxide, 21 % oxygen and 74 % nitrogen and the tests were limited to 10 min duration.

A further constraint arises because of the fact that, for routine clinical applications, any test signal should be simple in form and easy to apply. Thus the form of test input being applied involved switching of the inspired gas between dry air and the chosen gas mixture. Being binary in nature, this type of input requires only a hand-operated valve for its implementation. The period of switching between air and the gas mixture is under the control of the experimenter.

In general, all tests involved a period of 1 min of air breathing to allow the subject to become comfortable. After that initial interval, sampling of gas concentrations and flow rates (at a frequency of 10 Hz) was started and the chosen form of input switching pattern was applied. All channels of data were subjected to digital filtering with a simple low-pass filter having a breakpoint at 2.5 Hz.

It should be noted that the model equations include the instantaneous value of partial pressure of carbon dioxide in the inspired gas mixture as it enters the alveolar compartment ($P_I(t)$). What is measured, however, is the partial pressure of carbon dioxide at the mouth. Further information about the signal processing and parameter estimation processes carried out for this application may be found elsewhere [13, 14].

11.3.4 Experimental Design and Test Signal Selection

Measures of the quality of an experiment of the type being considered here may be based on the parameter information matrix, $\boldsymbol{M}$, and thus on the parameter sensitivity matrix, $\boldsymbol{X}$, which is itself dependent on the parameter values within the model. This means that the design of experiments for parameter estimation, or for the purposes of model validation, can never be optimal. The measures of quality can only be used to point the investigator in the right general direction in terms of experimental design and the selection of an appropriate test signal.

The approach used in developing test signals was based on simulation studies involving the nominal model and account was taken of the constraints discussed in the previous section in terms of the binary form of test input to be used, the concentrations of carbon dioxide permitted and the duration of the experiment. Switching periods considered in the simulation study ranged from 10 breaths to 70 breaths. Two different cases were considered. One involved finding the best test input to obtain estimates of all of the parameters of the model simultaneously from a single system identification test. The other involved finding optimal test inputs that could be applied for the estimation of each of the parameters individually and could be useful if it was believed that the uncertainty level for one parameter was particularly high and an experiment was required that would maximise the information content within the model response for that specific parameter.

The test input design process in the first case was based on use of the D-optimal criterion [15] which, as discussed in Chap. 5, involves minimisation of a cost function of the form:

$$J_D = \det\left(\boldsymbol{M}^{-1}\right) \tag{11.35}$$

and implicitly leads to information being distributed equally over all the parameters. For convenience, the measure of information content that was used involved the inverse of the cost function of (11.35). Results (Fig. 11.3) show that for a normal human subject under resting conditions the optimum switching period is about 55 breaths.

The second approach involved finding a test input that would allow information to be maximised in terms of specified parameters of the model using the truncated D-optimal design method [16] which involves the use of a cost function of the form:

$$J_{Dt} = \det\left(\boldsymbol{M}_{ii}^{-1}\right) \tag{11.36}$$

where $\boldsymbol{M}_{ii}$ is a sub-matrix of the full matrix $\boldsymbol{M}$ involving only the i parameters of interest. Again, the measure of information content that was used involved the inverse of the cost function.

Figures 11.4, 11.5, 11.6, and 11.7 show the results in this second case. For estimation of the parameters $\dot{M}$ and V_{TC} the test signal should involve long switching periods and thus a low frequency binary signal is useful. For the parameter representing the initial steady state volume of the alveolar compartment, $V_A(0)$, high frequencies of switching were found to be desirable and this is to be expected on physical grounds since this parameter is associated with a time constant which is small. The other two parameters are associated with the tissue compartment which has dynamics that are much slower than those of the alveolar compartment. The parameter $\dot{Q}$, which represents the total blood flow through the lungs (which is the same as the cardiac output in normal human subjects), was found to be estimated best using a test signal with a switching period in the lower part of the range considered, with an optimum at about 24 breaths.

It is interesting to note from the results of Figs. 11.3 and 11.4 that, for the cases that show a clear optimum (the case for the complete parameter set and the case which emphasises the information about the parameter $\dot{Q}$), there is an asymmetry in the curves which suggest the use of a switching period that is a little greater than the theoretical optimum in order to provide some robustness in the face of parametric uncertainties. Sensitivity studies, in which the sensitivity of the test input signal designs to variation of model parameters and the level of ventilation were

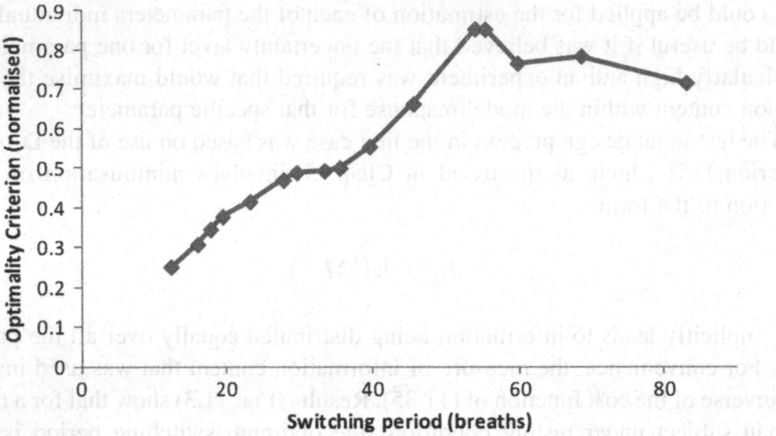

Fig 11.3 Test signal optimality investigation: results for estimation of the full parameter set

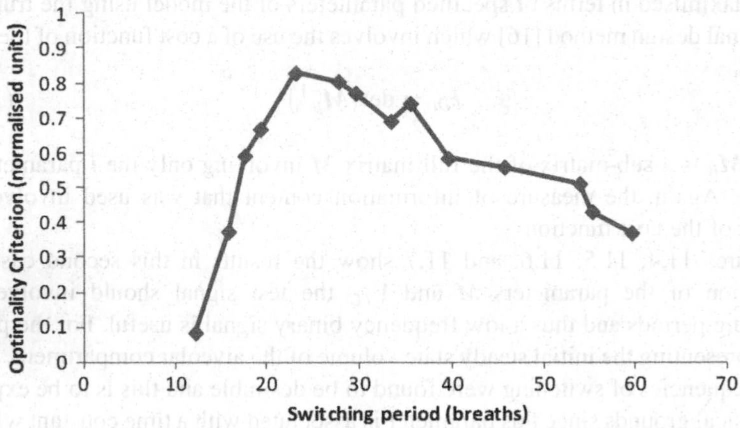

Fig. 11.4 Test signal optimality investigation: results for estimation of parameter $\dot{Q}$

investigated, suggest that the best test signals are relatively insensitive to variations within the normal physiological range under resting conditions.

The lung model was investigated through a series of tests involving human subjects whose lungs were assessed as being "normal" in terms of routine pulmonary function tests and, for these subjects, the model structure of Fig. 11.2 is considered appropriate. For cases in which parameters of the model had values found experimentally for the individual subjects using a maximum-likelihood method of system identification, validation results showed good agreement between measured and simulated outputs, with uncorrelated residuals [13]. It should be noted that, in the context of the test signal design findings discussed above, the

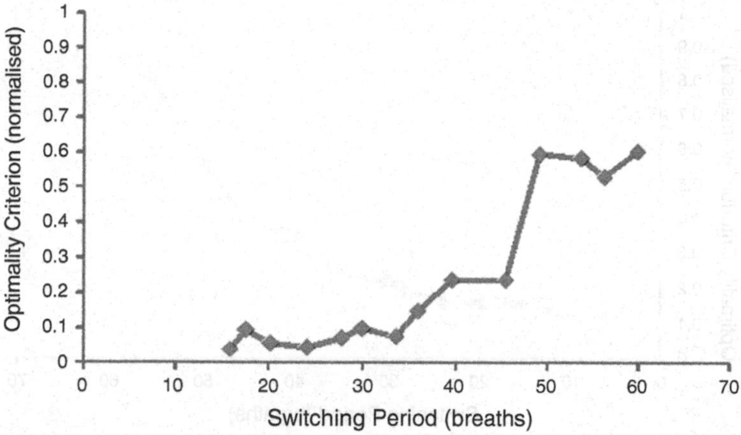

Fig. 11.5 Test signal optimality investigation: results for estimation of parameter $\dot{M}$

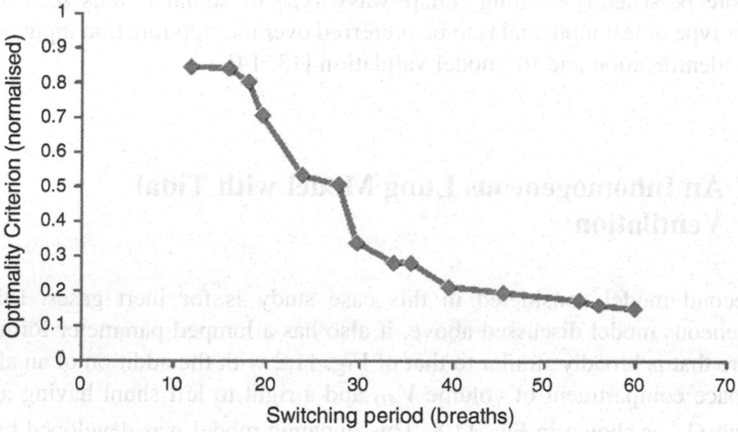

Fig. 11.6 Test signal optimality investigation: results for estimation of parameter V_A

switching period of the binary test signal used for model validation was 30 breaths in normal adult subjects (approximately 2 min).

In one specific, but typical, test the parameter correlation matrix corresponding to (11.34) was found to have relatively small off-diagonal elements, without any large interactions between the parameters being detected in the parameter correlation matrix. For example, for the parameters $\dot{M}$ and V_{TC} the value of the corresponding element in the matrix was found to be −0.017 (compared with 0.999 in the case of the step input), while the element showing interactions between $\dot{M}$ and V_A was 0.067 compared with 0.840 for results found using the step input. Use of the persistently exciting test signal thus avoids the problems of pathological unidentifiability that were encountered with the simple step function type of input used for the preliminary identification experiments and discussed in Sect. 11.3.1.

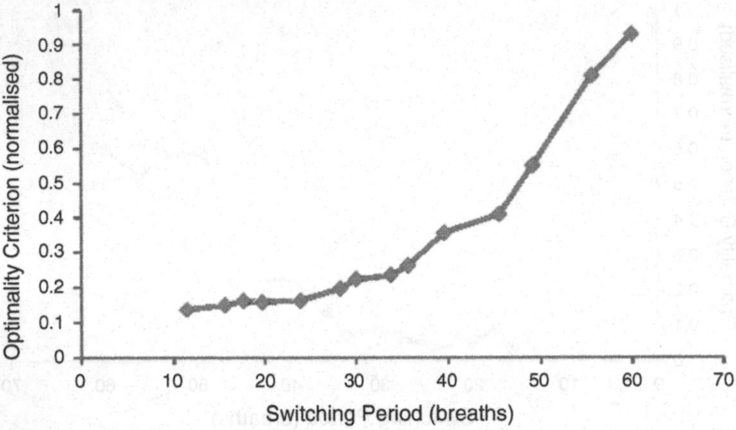

Fig. 11.7 Test signal optimality investigation: results for estimation of parameter V_{TC}

The more persistently-exciting square-wave type of signal is thus seen to be a superior type of test input and is to be preferred over the step function input both for system identification and for model validation [13, 14].

11.4 An Inhomogeneous Lung Model with Tidal Ventilation

The second model considered in this case study is for inert gases. Like the homogeneous model discussed above, it also has a lumped parameter form and a structure that is broadly similar to that of Fig. 11.2 with the addition of an alveolar dead-space compartment of volume V_{AD} and a right to left shunt having a blood flow rate $\dot{Q}_s$, as shown in Fig. 11.8. This dynamic model was developed by Pack [10] from an earlier steady-state model of Riley and Cournand [17].

Application of the principle of conservation of mass to the model structure of Fig. 11.8 leads to the following equations:

$$\frac{dP_A}{dt} = \frac{Sk\dot{V}}{V_A}(P_I^* - P_{AD}) + \frac{k_1\dot{Q}\alpha_{BL}}{V_A}[(P_{TC} - P_A)] \tag{11.37}$$

$$\frac{dP_{AD}}{dt} = \frac{S(1-k)\dot{V}}{V_{AD}}(P_I^* - P_{AD}) \tag{11.38}$$

$$\frac{dP_{TC}}{dt} = \frac{\dot{Q}\ k_1}{V_{TC}}[P_a - P_{TC}] \tag{11.39}$$

where k represents the fraction of the total ventilation associated with the ideal alveolar compartment and k_1 is the fraction of the total blood flow $\dot{Q}$ that is

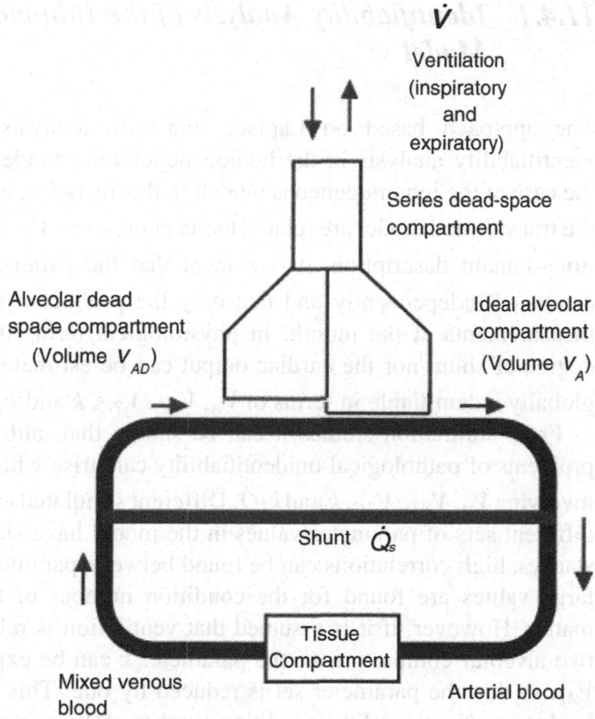

Fig. 11.8 Structure of inhomogeneous model of gas-exchange processes

associated with the ideal alveolar compartment. The remainder ($\dot{Q}_S$) of the cardiac output passes through the shunt pathway. The quantity a_{BL} is the Ostwald coefficient between the lungs and the blood.

The partial pressure of the chosen inert gas within the ideal alveolar compartment is represented by $P_A(t)$, while $P_{AD}(t)$ is the partial pressure of that gas within the alveolar dead-space compartment. The arterial partial pressure, $P_a(t)$ for this model structure is given by:

$$P_a(t) = k_1 P_A + (1 - k_1) P_{TC} \tag{11.40}$$

The volumes of the alveolar compartments, V_A and V_{AD}, are assumed to vary with time in proportion to their ventilator flow rates. During Stage 1 of the breath cycle it is assumed that the gas leaving the common dead space compartment has a partial pressure value equal to the flow-weighted mean of the partial pressures values of the gases from the two alveolar compartments over the end-tidal part of the previous expiration.

The output variable for the inhomogeneous model is defined as the flow-weighted sum of the partial pressures of gas in the two alveolar compartments. This corresponds to the measured partial pressure of gas at the mouth, provided account is taken of the time delay introduced within the dead space.

11.4.1 Identifiability Analysis of the Inhomogeneous Lung Model

The approach based on Laplace transform analysis which was adopted for identifiability analysis in the homogeneous lung model can again be applied for the case of the inhomogeneous model. In this instance, when the coefficients within the transformed model are related to the parameters V_A, V_{AD}, V_{TC}, $\dot{Q}$, k and k_1 of the time-domain description, it is evident that the parameters k_1 and $\dot{Q}$ cannot be estimated independently and that only the product $k_1\dot{Q}$ may be estimated from measurements at the mouth. In physiological terms this shows that neither the degree of shunt nor the cardiac output can be estimated. The model is, however, globally indentifiable in terms of V_A, V_{AD}, V_{TC}, k and $k_1\dot{Q}$.

From simulation studies it can be shown that, although globally identifiable, problems of pathological unidentifiability can arise with the reduced parameter set involving V_A, V_{AD}, V_{TC}, k and $k_1\dot{Q}$. Different simulated experimental conditions and different sets of parameter values in the model have shown that, in some circumstances, high correlations can be found between parameters and that in these cases large values are found for the condition number of the parameter information matrix. However, if it is assumed that ventilation is related to the volumes of the two alveolar compartments, the parameter k can be expressed in terms of V_A and V_{AD} so that the parameter set is reduced by one. This change of model structure leads to a reduction of the condition number of the parameter information matrix by a factor of at least ten and thus improves the situation significantly without adversely affecting the global identifiability. However, in contrast with the results for the homogeneous model, the results for the inhomogeneous model suggest that estimation of parameters is impossible without additional measured variables [18].

11.5 Applications of Respiratory Gas-Exchange Models

Models of human respiratory gas exchange processes are important for a number of reasons. They provide sub-models for use within more general models of the human respiratory control system and many different models have been proposed for this complex physiological system. Early experimental investigations include those by Gray [19] and by Lloyd and Cunningham [20], together with model-based studies such as those by Grodins et al. (see, e.g. [21, 22]). Later modelling studies such as those of Saunders et al. [23], Tomlinson et al. [24] and Lin et al. [25] have included more functional components and provided more insight in a number of different respects. In some cases these models have incorporated the type of dynamic gas exchange representation that involves tidal ventilation (as presented in Sect. 11.3.1). Such tidal ventilation models are particularly important for respiratory control studies because of the fact that the inspiratory and expiratory phases of the breath cycle are represented explicitly. Some recent studies have been concerned more specifically with modelling the system during exercise (see e.g., [26]) and in sleep-disordered breathing (see e.g. [27]).

Gas exchange models also provide a basis for tests of lung function and a useful review of methods of clinical measurement in respiratory medicine has been provided by Flenley and Warren [2]. That review includes a number of sections, such as those dealing with gas exchange between alveoli and capillaries, the carriage of oxygen by the blood and also the transport of carbon dioxide, in which explicit use is made of gas exchange models. It is clear that restrictions posed by methods of lung function testing which depend upon a steady-state being achieved in the patient can, in some cases, be improved upon by the use of dynamic models of gas exchange processes. For example, it has been suggested [18] that system identification techniques could provide a basis for discrimination between homogeneous and inhomogeneous lung models and thus offer the possibility of estimating any maldistribution of ventilation in the subject under investigation. A hypothesis testing approach of the kind proposed by Goodwin and Payne [28] has been suggested for use with the inhomogeneous lung model. In this a test of the null hypothesis that the volume of the alveolar dead space compartment (V_{AD}) is zero is equivalent to testing the hypothesis that the homogeneous inert gas model is preferred to the inhomogeneous model. This could perhaps allow the relatively simple compartmental models outlined in this case study to be used to distinguish between subjects for whom a homogeneous description is appropriate and others who have some form of maldistribution of ventilation. In addition to testing the suitability of different model structures for individual subjects this could, possibly, allow a distinction to be made between "normal" and "abnormal" lungs at a clinically useful stage in the development of some form of pulmonary or circulatory disorder [14].

The estimation of the pulmonary blood flow is important for many clinical purposes and many methods have been developed for this. The "indicator dilution" method, which is an invasive procedure, is the approach which is used most widely. However, there is considerable interest in the development of less invasive techniques for the estimation of pulmonary blood flow. The use of measurements of gas partial pressures and flow rates at the mouth in conjunction with dynamic models of the gas exchange processes has provided a basis for research on a number of approaches. Ideas relating to the estimation of cardiac output from measurements of inhaled and exhaled blood-soluble anaesthetic gas can be traced back to the work of Fick in 1870 [29]. Early work on this type of approach involving the type of gas exchange model for carbon dioxide with tidal ventilation described in Sect. 11.3.1 was reported by Ferguson et al. [9] and developed further by Pack [10] and by Bache et al. [13]. The identifiability analysis presented in Sect. 11.3.2 and the work reported in Sect. 11.3.4 were performed initially with this objective in mind, as well as the non-invasive estimation of other parameters within gas exchange models [13]. In other research work of a similar kind Brovko et al. [30] applied an approach based on the extended Kalman filter with the aim of tracking changes of pulmonary blood flow. Examples of other relevant research involving the use of system identification methods for estimation of pulmonary blood flow include the work of Watanabe et al. [31], Akman et al. [32] and Noshiro et al. [33]. In several of those studies considerable emphasis has been placed on the importance of the length of experimental record in terms of the accuracy of estimates, and on the need for objective tests of the identified model. It is interesting to note that in the investigations of Bache et al. [13] and Noshiro et al. [33] quantitative tests of model adequacy were based on correlation functions of the residuals between the measured output and corresponding model variable, as discussed in Chap. 7. It should be noted that, in recent years, developments in technology have led to more research on methods of Doppler electrocardiography and thoracic electrical bio-impedance methods to provide minimally – invasive methods for pulmonary blood flow estimation. Commercially available equipment now allows for routine monitoring of pulmonary blood flow using these ultrasonic and electrical methods of measurement but there is also equipment commercially available which is based on parameter estimation methods using a gas exchange model for carbon dioxide re-breathing (see e.g. [34]).

One important area for the application of dynamic gas exchange models has been in the development of teaching aids which can be used to illustrate concepts in respiratory physiology and medicine. Experimentation with a realistic computer simulation, equipped with a robust and effective user interface, can be especially effective in developing an understanding of the complex interactions that can be found within a system of this kind. Examples of publications describing the successful application of simulation methods applied to the teaching of respiratory physiology include simulation models specifically describing pulmonary gas exchange concepts (see e.g. [35, 36]) and more general simulation models of the respiratory system such as the MACPUF model developed by Dickenson et al. [37–39].

11.6 Quality Issues in Gas-Exchange Modelling

Much of the discussion in the previous sections of this chapter has related to experimental modelling methods although based on a model structure derived from the underlying anatomy, physiology and physics of the system. In such an approach, issues of model quality can be related directly to questions of identifiability and experimental design, as in Sects. 11.3.2 and 11.3.4. Clearly such issues are of fundamental importance for the application of lung gas exchange models to individual subjects, where the model structure and parameter values need to be appropriate for that subject. For applications relating to an individual subject, such as those concerned with non-invasive methods for estimation of quantities such as the pulmonary blood flow or lung volume, model validation tests must be carried out for each model generated for that subject.

The fitting of a model to an individual subject is not always essential. For some types of research application all that may be needed is a model of gas exchange processes that is correct in that it has been developed using appropriate scientific principles, with a structure and parameters that are "average" for a specific group of human subjects. For example, this "average" gas exchange model may then be incorporated into a larger model, for investigation of some aspect of the respiratory control system. Validation in situations of this kind most often involves expert opinion and peer review.

It is interesting to note that, in recent years, some research groups have included full details of the equations, structure and parameters of simulation models within their published work. Examples may be found in the papers of Cheng et al. [27] and Hardman et al. [40], where details of models and parameter values are included in full. Details of the gas exchange model with tidal ventilation described in Sect. 11.3.1 may be found within the information included on the website that is associated with this book. Some basic physiological background is also provided, along with links to other sources of information on gas-exchange models and their applications in respiratory physiology and medicine.

References

1. Petersen ES, Cunningham DJC (1988) Respiratory physiology. In: Cramp DG, Carson ER (eds) The respiratory system, vol 2, Measurement in Medicine Series. Croom Helm, London, pp 13–96
2. Flenley DC, Warren PM (1988) Clinical measurement in respiratory medicine. In: Cramp DG, Carson ER (eds) The respiratory system, vol 2, Measurement in Medicine Series. Croom Helm, London, pp 97–144
3. Herczynski R (1988) The mathematical study of respiratory phenomena. In: Cramp DG, Carson ER (eds) The respiratory system, vol 2, Measurement in medicine series. Croom Helm, London, pp 145–295
4. Murray-Smith DJ, Carson ER (1988) The modelling process in respiratory medicine. In: Cramp DG, Carson ER (eds) The respiratory system, vol 2, Measurement in medicine series. Croom Helm, London, pp 296–333
5. Murray-Smith DJ, Carson ER (1988) Case studies in respiratory system models. In: Cramp DG, Carson ER (eds) The respiratory system, vol 2, Measurement in medicine series. Croom Helm, London, pp 334–389
6. Ben-Tal A (2006) Simplified models for gas exchange in the human lung. J Theor Biol 238:474–495
7. Hahn CEW, Farmery AD (2003) Gas exchange modelling: no more gills please. Br J Anaesth 91(1):2–15
8. Murphy TW (1969) Modelling of lung gas exchange – mathematical models of the lung. Math Biosci 5:427–447
9. Ferguson DR, Mills RJ, Moran F et al (1973). Estimation of the parameters of the lung with clinical applications. In: Proceedings of the 3rd IFAC symposium on Identification and system parameter estimation, North-Holland, pp 213–219
10. Pack AI (1976) Mathematical models of lung function. PhD dissertation, University of Glasgow, UK
11. Piiper J, Dejours P, Haab P et al (1971) Concepts and basic quantities in gas exchange physiology. Respir Physiol 13:293–304
12. Peters JP, Bulger HA, Eisenman AJ (1924) Studies of the carbon dioxide absorption curve of human blood IV. J Biol Chem 58:47–768

13. Bache RA, Gray WM, Murray-Smith DJ (1981) Time-domain system identification applied to non-invasive estimation of cardio-pulmonary quantities. IEE Proceedings, 128, Part D:56–64
14. Bache RA (1981) Time-domain system identification applied to non-invasive estimation of cardio-pulmonary quantities. PhD dissertation University of Glasgow, UK
15. Federov VV (1972) Theory of optimal experiments. Academic, New York
16. Hunter WG, Hill WJ, Henson TL (1969) Designing experiments for precise estimation of some of the constants in a mechanistic model. Can J Chem Eng 47:76–80
17. Riley RL, Cournand A (1951) Analysis of factors affecting partial pressures of oxygen and carbon dioxide in gas and blood of lungs: theory. J Appl Physiol 4:77–103
18. Bache RA, Murray-Smith DJ (1983) Structural and parameter identification of two lung gas-exchange models. In: Vansteenkiste GC, Young PC (eds) Modelling and data analysis in biotechnology and medical engineering. North-Holland, Amsterdam, pp 175–188
19. Gray JS (1946) The multiple factor theory of the control of ventilation. Science 103 (2687):739–744
20. Lloyd BB, Cunningham DJC (1963) A quantitative approach to the regulation of human respiration. In: Cunningham DJC, Lloyd BB (eds) The regulation of human respiration. Blackwell, Oxford, pp 331–349
21. Grodins FS, Gray JS, Schroeder KR et al (1954) Respiratory responses to CO_2 inhalation: a theoretical study of a nonlinear biological regulatory. J Appl Physiol 7:283–308
22. Grodins FS, Buell J, Bart AJ (1967) Mathematical analysis and digital simulation of the respiratory control system. J Appl Physiol 22(2):260–276
23. Saunders KR, Bali HN, Carson ER (1980) A breathing model of the respiratory system: the controlled system. J Theor Biol 84:135–161
24. Tomlinson SP, Tilley DG, Burrows CR (1994) Computer simulation of the human breathing process. IEEE Eng Med Biol 13:115–124
25. Lin S-L, Guo N-R, Chiu C-C (2012) Modeling and simulation of respiratory control with LabVIEW. J Med Biol Eng 32(1):51–60
26. Thamrin H, Murray-Smith DJ (2009) A mathematical model of the human respiratory control system during exercise. In: IASTED proceedings 659 (IASTED International conference on modeling simulation and identification, Oct 12–14 2009, Beijing, China), ACTA Press, Calgary. Online: www.eprints.gla.ac.uk/66888/. Accessed 13 Aug 2015
27. Cheng L, Ivanova O, Fan H-H et al (2010) An integrative model of respiratory and cardiovascular control in sleep-disordered breathing. Respir Physiol Neurobiol 174:4–28
28. Goodwin GC, Payne RL (1977) Dynamic system identification: experiment design and data analysis. Academic, New York
29. Fick A (1870) Űber die Messung des Blutquantums in den Herzventrickeln, Sitzungsberichte der physikalischmedizinischen Gesellschaft zu Wűrzburg, 16
30. Brovko O, Wiberg DM, Arena L et al (1981) The extended Kalman filter as a pulmonary blood flow estimator. Automatica 17:213–220
31. Watanabe K, Sekiguchi H, Uchiyama A et al (1983) IECE Jpn Trans E66(4):254–255
32. Akman G, Kaufman H, Roy R (1985) Estimation of cardiac output using quasi-optimal inputs and an approximated maximum-likelihood algorithm. In: Barker HA, Young PC (eds) Identification and system parameter estimation. North Holland, Amsterdam, pp 1203–1208
33. Noshiro M, Maekawa H, Ishida A et al (1988) Accuracy in noninvasive estimation of pulmonary blood flow by identification of the CO_2 uptake system. Front Med Biol Eng 1 (1):45–58
34. Funk DJ, Moretti EW, Gan TJ (2009) Minimally invasive cardiac output monitoring in the perioperative setting. Anesth Analg 108(3):887–897
35. Mills RJ, Middleton S, Moran F et al (1974) Simulation in the teaching of concepts of respiratory gas exchange. Int J Math Educ Sci Technol 5:381–387
36. Kapitan KS (2008) Teaching pulmonary gas exchange physiology using computer modelling. Adv Physiol Educ 32(1):61–64

37. Dickinson CJ (1972) A digital computer model to teach and study gas transport and exchange between lungs, blood and tissues ("MACPUF"). J Physiol 224:7P–9P
38. Dickinson CJ (1977) A computer model of human respiration. University Park Press, Baltimore
39. International Network for Humane Education (InterNICHE), Mac series of physiological and medical simulations. Online: www.interniche.org/it/node/475. Accessed 01/02/2015
40. Hardman JG, Bedforth NM, Ahmed AB et al (1998) A physiological simulator: validation of its respiratory components and its ability to predict the patient's response to changes in mechanical ventilation. Br J Anaesth 81:327–332

Chapter 12
Case Study: Modelling of Elements of the Neuromuscular Systems Involved in Regulation of Posture and Control of Movement

12.1 Introduction

Neuromuscular systems involve elements of the muscular and nervous systems that are central to a range of different physiological functions, including the regulation of posture and control of movement, the control of breathing and the operation of the heart. This case study is concerned with the first of these physiological functions – control of posture and movement.

The neural and muscular systems involved in the movement of limbs and the control of posture are, very complex, hierarchical and highly nonlinear. They also present major problems in terms of experimental investigations. The observation of quantities such as body or limb position and velocity, forces exerted by the limbs and electro-myographic activity is relatively straightforward but measurements of other, internal, quantities is challenging. Experimental assessment of what is happening within the highly parallel structures of the system is, in many cases, impossible with currently available technology.

For decades these systems have attracted the attention of researchers from many fields, in addition to physiology. Computer scientists, engineers, physicists and applied mathematicians have all made important contributions. This interest is due in part to the potential clinical importance of improving our understanding of these systems and also to the fact that neuromuscular control systems have features that are relevant for some engineering applications, such as robotics. Concepts borrowed from the field of control engineering are now, in turn, being used in physiology to provide a fuller understanding of some of the subtleties of these complex electro-mechanical biological systems. Inevitably, mathematical and computer-based models are central to these interdisciplinary exchanges and model quality and model validation issues are therefore of great importance.

Although successful to some extent, it is generally accepted that multidisciplinary model-based investigations have not yet provided answers to many

D.J. Murray-Smith, *Testing and Validation of Computer Simulation Models*,
Simulation Foundations, Methods and Applications,
DOI 10.1007/978-3-319-15099-4_12

questions about the basic principles of these biological systems and, in particular, how they allow us to achieve the extraordinary levels of performance that we all take for granted in everyday activities such a walking, running and lifting as well as in more complex tasks. The growth of rehabilitation engineering, prosthetics and robotics has given rise to new questions and to new areas of research, but many important gaps remain in our basic understanding.

As well as providing an example of a biomedical application of simulation modelling techniques, the use of this physiological system for a case study is also attractive because of the fact that the system is essentially hybrid. It involves an interesting mix of continuous variables (such as muscle length and muscle tension) and inherently discrete variables (such as those involved in the communication pathways of the nervous system). The problems of validating models such as this are challenging because of the highly complex behaviour patterns exhibited by the real system, the parallel structures found within the system, the inevitable simplifying assumptions needed for a mathematical or computer-based representation, the issues of inherent biological variability and, most importantly, the major uncertainties that still exist about how the system functions.

This case study has been developed from material presented some years ago as part of a course on simulation and modelling techniques at Master's degree level presented, primarily, to graduate students from the biological sciences but also involving some who had a mathematical or physical sciences background [1]. There are significant challenges in the preparation of a case study of this type since readers are unlikely to have all the background knowledge to be able to understand the experimental limitations, the methods used for investigation and the basic uncertainties about how the system works. Those with a background in computer science or engineering are likely to have rather limited knowledge of physiology and may require access to additional background material to establish the necessary foundations. Conversely, those with backgrounds in biology may require some support to guide them through the relevant fundamentals in terms of mathematics, computer simulation and control engineering concepts. Support material, especially in terms of lists of directly relevant and easily accessible reference books, research reports and published review articles, may be identified through the web-based resources for this book to allow readers of this chapter to dig more deeply into this topic.

12.2 Mathematical and Computer-Based Modelling of the System

Although some physiologists still express doubts about the contribution made by modelling techniques and computer simulation methods to our current understanding of the neuromuscular system, many would now accept that models have been of some value in providing tools to assist in the development of hypotheses and for the design of experiments for hypothesis testing as part of the scientific method. Most would agree that modelling has a potentially important role in this field, provided the use of models is part of a combined theoretical and experimental programme of research and provided model testing always forms a central element.

Modelling and simulation methods have been applied at many different levels in neurophysiology, ranging from models of single nerve cells to attempts to represent, in a functional way, the complete multi-loop hierarchical systems controlling movement. For example, one classic investigation which led to a much improved understanding of the basic processes of nerve conduction was the work by Hodgkin and Huxley [2]. Their models of nerve cell membranes and axons established a method of analysis that has since been applied to a wide range of related problems and established an experimental approach that has proved to be of fundamental importance. The widely used Hodgkin-Huxley model, published first in 1952, is still attracting attention in terms of the modelling issues and especially the assumptions inherent in the model, its validation and the accuracy of its quantitative predictions [3]. Models have also been developed for many other elements of the system, leading to the development of models of complete stretch reflexes and the hierarchical structures involved in postural control and movement. Recent research involving the testing of hypotheses about mechanisms involved in bipedal locomotion is one interesting example (see, e.g. Prochazka and Gillard [4]).

The difficulty of developing and applying quantitative modelling techniques lies not only in the inherent complexity of these systems but also in the problems associated with the use of experimental modelling methods. The system has a highly redundant structure, with many parallel neural pathways and many sense organs that generate feedback signals, only a few of which can be monitored in any single experiment. This presents obvious difficulties for model development and model validation, since experiments are generally hard to perform and data collected in one experiment may be insufficient to allow proper validation as well as system identification. Issues of validation are returned to in several later sections within this case study.

12.3 The Physiology of Neuromuscular Systems

As already implied in the sections above, the neuromuscular systems that are involved in the control of limbs and the regulation of posture are hierarchical in structure and may be split conveniently into central and peripheral elements. The peripheral nervous system involves a sequence of repeating units within the spinal cord, known as "segmental" levels. At each segmental level of the spinal cord the elements of the peripheral neuromuscular system are as shown in Fig. 12.1 and involve the muscular and skeletal sub-systems, together with sensory receptors. These, together with the spinal cord segment, all lie within feedback pathways that form a stretch reflex system. The higher levels of this hierarchical structure involve elements of the central nervous system associated with the brain. Connections

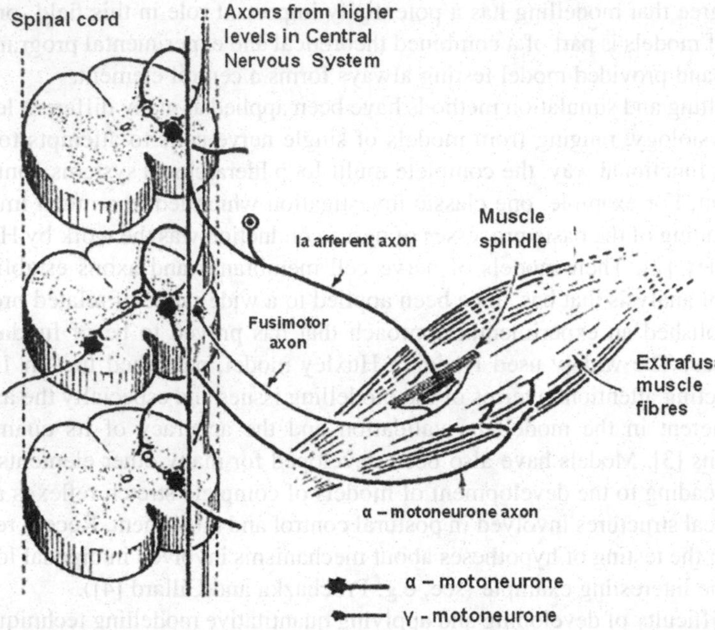

Fig. 12.1 Schematic diagram showing some of the elements of the neuromuscular system at one segmental level of the spinal cord. Signal transmission pathways connecting a muscle spindle sensory receptor and the spinal cord are shown, together with some neuronal circuits within the spinal cord (From Rosenberg et al. [33])

between elements of the system are made through communication pathways involving nerve cells.

Actuators within the stretch reflex system are the load-bearing or extrafusal muscle fiber which can contract in response to nervous stimuli. A single anatomical muscle contains a large number of contractile elements, which are organised in groups to form motor units that are connected in parallel to a tendon and thus exert force on a limb. Each motor unit typically involves about 250 extrafusal fiber and has a single nerve fiber (the motoneurone). The extrafusal muscle fiber contain smaller elements known as fibrils which themselves contain a large number of microfilaments. These microfilaments are either thin filaments of actin or thick filaments of myosin and the active contractile properties are associated with the movement of one of these types of filament over the other. Extrafusal fiber can be regarded as nonlinear springs which have the important additional property of being able to develop force in an active fashion. Thus, unlike the case of a simple mechanical spring, forces developed in muscle depend not only on external loading but also on the neural input from the motoneurone.

As already indicated, the sensory receptors act within local feedback pathways to provide reflex action. In highly simplified terms, this means that any stretching of extrafusal muscle gives rise to more activity in that muscle due to the sensory feedback and this makes the muscle fiber contract more strongly to counter the effect of the initial stretching. Skeletal muscle thus resists external forces in two distinct ways, partly through the stiffness of muscle itself and partly as a result of reflex action involving changes in activation level of the muscle fiber. The action of the elements within these reflex pathways may be influenced by activity at the next level above this in the nervous system. For example, one particular type of sensory receptor, known as a "muscle spindle", has neural inputs as well as neural outputs and these inputs can modify the behaviour of a muscle spindle as a sensory receptor. Central pattern generators at this higher level within the nervous system also provide neural inputs to the reflex system for repetitive actions such as walking. Interactions between different loops associated with different muscles are known to take place and this is important for coordination of activity in the different muscles that act at a joint. The higher elements within the hierarchical structure, such as the brain stem, cerebellum, motor cortex and sensory cortex are believed to deal with the planning and initiation of limb movements, with ascending and descending neural pathways providing communication links between the different levels.

Understanding of these systems tends to be greatest at the level of the peripheral nervous system, where experimentation is most straightforward. Useful review papers have been written about the stretch reflex system and the associated sensory receptors (e.g. [5–7]) and about the neuromuscular control more generally (e.g. [4, 8–10]). In contrast, much less is known about how the brain controls and coordinates muscle activity. There is also relatively little quantitative understanding of dynamic interactions between different reflex systems such as the stretch reflex and head and neck reflexes.

One well-established theory is that the neuromuscular control system in humans and other animals operates in an open-loop feed-forward fashion for fast movements, using stored muscle activation patterns (see, e.g. [11]) and that feedback is only significant for slower movements and for postural regulation. It has also been suggested that the system involves some combination of open-loop control and feedback control, but with switching operations that occur in addition to continuous feedback control (see, e.g. [9]). Issues of this kind are the subject of further discussion in Sect. 12.6.

12.4 Models of Skeletal Muscle

Skeletal muscles are active systems that convert chemical energy into mechanical energy and exert active force by contraction in response to a nervous stimulus. Muscles are normally organised in pairs, or in groups that allow directional movement about a joint. Muscles have two distinct states which are the resting (non-contracting) state and the active (contracting) state during which mechanical energy is produced. An electrically polarised membrane surrounds each fiber of an anatomical muscle and if this membrane is depolarised, transiently, contraction occurs within the muscle fiber. Within the intact system this depolarisation process is initiated when a nerve impulse in the motor nerve leading to the muscle reaches a junction area known as an end-plate. A wave of depolarisation is then transmitted along the fiber and this gives rise to a single twitch response. Repetition of the stimulus before the mechanical effect of the first response has died away leads to summation of the mechanical responses. If the stimulus is applied at increasing frequencies the twitch responses tend to merge producing a steady contraction with a small fluctuation at the stimulus frequency superimposed on it. The mean level of contraction depends on the frequency of the neural stimulus. Eventually, as the stimulus frequency is further increased, a condition occurs in which there is simply a steady contraction without any superimposed fluctuations. This condition is termed "tetanus" and increasing the stimulus frequency beyond this limit produces no further change of output.

Modelling of skeletal muscle has been carried out for many different purposes and at many different levels of detail, ranging from microscopic models which reflect current understanding of the physical and chemical mechanisms of contraction within individual muscle fiber to forms of empirical model describing the overall properties of complete muscles which are derived entirely on an experimental basis [12–14]. The highly detailed models, although of great value to physiologists in providing a theoretical basis for contractile mechanisms, are not appropriate as a basis for modelling the complete neuromuscular control system. For applications of that kind a simpler and more empirical type of model is more appropriate.

12.4.1 Development of an Experimental Model of Skeletal Muscle

No single experimentally-based model has been developed that describes, in a comprehensive fashion, the behaviour of skeletal muscle for a wide range of loading conditions. Most published models have been based on measurements and observations from a single type of experiment. The example considered in this case study involves a model of muscle which was developed for use in neuromuscular control system investigations [15]. Like many other models, it has an empirical basis, but experience gained during its development has also allowed it also to be used successfully as an element within models of the stretch reflex system and as a starting point for models of sensory receptors. The model is based on experimental data gathered over half a century ago and has been chosen as the basis of this case study because it is relatively simple and, hopefully, understandable by those who have little biological knowledge. It does, however, illustrate well the processes of model development and testing in the biomedical field and the types of difficulty that can arise.

A useful starting point in the development of experimentally based models of muscle involves a lumped-parameter analogue having three mechanical elements.

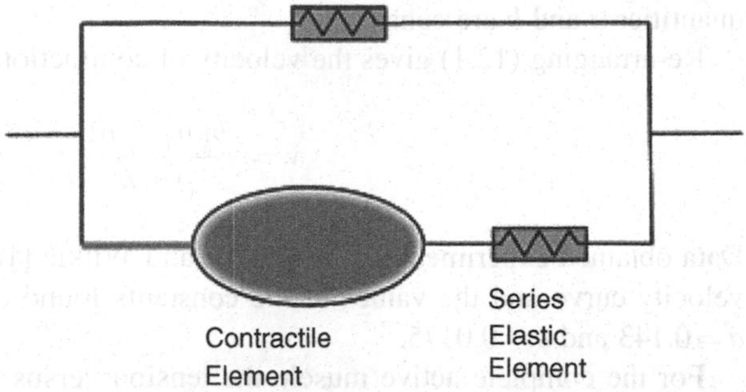

Fig. 12.2 Schematic diagram showing the three elements of the mechanical model of skeletal muscle

This mechanical analogue was formulated by Hill [16] in the early 1920s. His experimental investigations had suggested that active skeletal muscle could be represented by a system of lumped-parameter elements having mechanical properties that could be described in terms of elasticity and viscosity [16, 17]. The elastic elements are defined as those in which tension is a function of muscle length while, in the viscous elements, tension is a function of velocity. Hill's model contains an un-damped elastic element (known as the "series-elastic" element) together with an element in series with this that has properties that involve damping (known as the "contractile" element). Resting muscle has elastic characteristics but only has a small amount of damping and Hill's experimental studies also suggested the presence of a second elastic element (known as the "parallel elastic" element) in parallel with the series elastic and contractile elements, as shown in Fig. 12.2.

For the purposes of characterising a muscle in terms of this very simple three element conceptual description, several experimentally-derived curves are needed [18]. These include a force-length curve and a force-velocity curve for the complete active muscle, a force-extension curve for the series elastic element and a force-length curve for the parallel elastic element. A curve is also needed to describe the muscle activity, as a function of time, following the arrival of a motor nerve impulse at the muscle end-plate. This curve, which defines the temporal properties of active muscle, is termed the "active-state" curve. For conditions in which the muscle length is much less than the normal resting length the stiffness of the parallel elastic element can usually be neglected leaving a simpler two-element series type of model.

The force-velocity curve is of special importance and has been investigated in particular detail. A rectangular hyperbola is known to provide a good fit to experimental results for most skeletal muscles during active contraction. This is known as Hill's equation and is normally written in the form:

$$(p+a)(v+b) = (p_0+a)b \tag{12.1}$$

Here the variable p represents the force developed by the muscle, v is the speed of shortening of the muscle, p_0 is the muscle tension when the velocity is zero and the quantities a and b are constants.

Re-arranging (12.1) gives the velocity of contraction of active muscle as:

$$v = \frac{b(p_0 - p)}{p + a} \tag{12.2}$$

Data obtained experimentally by Ritchie and Wilkie [18], were used to fit a force-velocity curve and the values of the constants found empirically in this way are $a = 0.143$ and $b = 0.0175$.

For the complete active muscle the tension versus length curve can be found experimentally by imposing tetanic conditions on the muscle when it is constrained so that its length remains constant (isometric conditions). The overall muscle tension is measured for several sets of length conditions and this allows an approximate force-length curve to be fitted [18] which can be represented by the equation:

$$p = p_c - k(l - l_0)^2 \text{ for } l < l_0. \tag{12.3}$$

Here, as in (12.1), the variable p is the muscle force, l is the length of the muscle, l_0 is the normal length of the muscle and p_c is the muscle tension at muscle length l_0. The values of the constants found in this way are $k = 2060$, $p_c = 0.481$ and $l_0 = 0.032$. It should be noted that experimental data are not available from [18] for cases in which the muscle length l is greater than l_0.

The length of the series-elastic element cannot be determined experimentally and indirect measurement methods must be used to characterise that element of the model [18]. Through this approach it is possible to relate the extension of this element to the force developed within it and establish a look-up table that represents the characteristics of this highly nonlinear element.

The active-state curve relates the degree of activity of the contractile element to time, following the application of a stimulus from the nervous system. Although there are uncertainties about the precise form of the initial rising phase of the curve, it is known that this is very rapid in comparison with the dynamics of muscle and it may be assumed that the element becomes an active source of tension immediately after the arrival of the nerve impulse at the end-plate. The tension then remains constant for an interval before starting to fall gradually towards zero. Figure 12.3 is the curve used in this model to show how the active state, h, varies with time following the arrival of a nerve impulse at time $t = 0$. In this representation it has been assumed that the initial rise is instantaneous. In terms of the active state curve representation, the activity at any time instant following activation is expressed as a fraction of the maximum and the tension p_0 at zero velocity for a given muscle length l is described by the equation:

$$p_0 = h(t)[p_c - k(l - l_0)]^2 \tag{12.4}$$

where $h(t)$ represents the normalised values found from the active-state curve for each instant of time following the application of the stimulus.

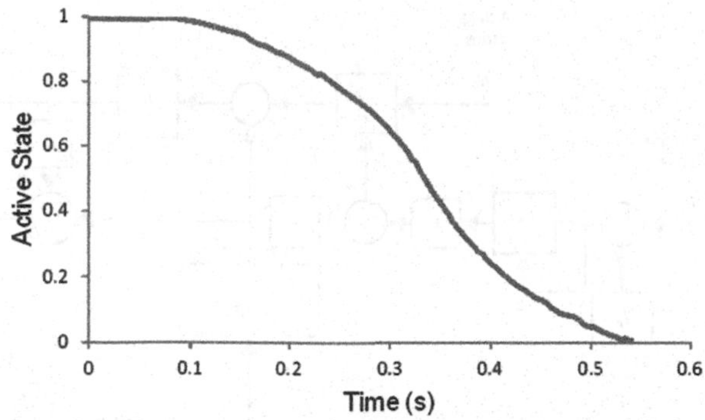

Fig. 12.3 Plot of variation of active state of muscle with time following a neural stimulus. This provided the basis for the active state function used within the muscle model (Data for this function were derived from Ritchie and Wilkie [18])

Under isometric conditions the overall length of the muscle remains constant and, from (12.2), we can establish the tension at zero velocity for the given length, assuming the muscle activity to be at the maximum. Any change of length of the contractile element must result in an equal (and opposite) length change in the series elastic element. This length change, Δl, produces a tension change and this must be the overall force, p, developed within the model. Figure 12.4 shows a block diagram for this muscle model for these isometric conditions.

For isotonic conditions the overall force within the muscle is controlled externally and the force developed by the contractile element and the force within the series elastic element must always equal that external force. An increase in neural activity causes the contractile element to shorten, but the series elastic element cannot change its length because that would result in a change of tension which is not possible because of the isotonic requirement. The change of length of the complete muscle must therefore be equal to the change of length of the contractile element. Figure 12.5 shows the block diagram for the isotonic case.

12.4.2 Testing and Validation of the Muscle Model

Available test results for the quantitative validation of the model based on the representation outlined above are very limited. The only immediately useful results involve data from a series of experiments for isotonic conditions performed by Ritchie and Wilkie [18] at the same time as the tests carried out to establish the parameters of the model. It should be noted that in these isotonic experiments, the muscle is held initially under isometric conditions until the active tension due to the applied nervous stimulus is equal to the applied external force and the muscle then starts to contract. This is known as an after-loaded twitch response. The parameter estimation processes used in the development of the model did not involve tests of that kind and this means that the after-loaded twitch response experiment is an appropriate case to consider for model validation purposes.

Figure 12.6 represents the after-loaded isotonic condition. In this case the muscle is allowed to develop tension while held at constant length until the muscle force reaches the required value. From that moment isotonic conditions are imposed, with the tension being held constant by external means. The block diagram model in this

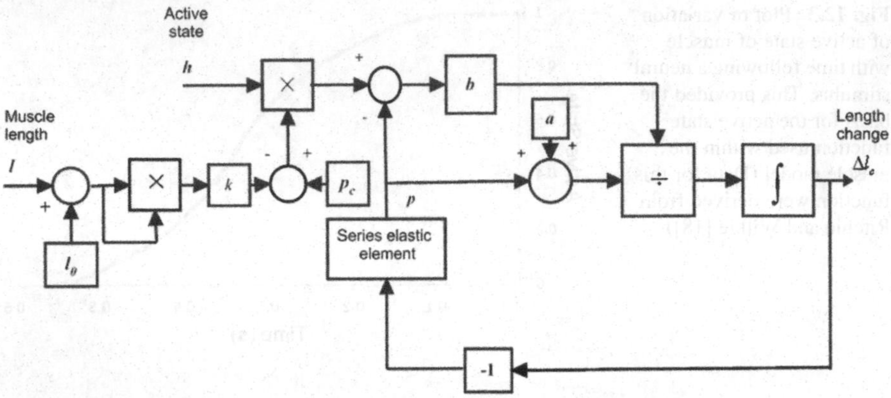

Fig. 12.4 Block diagram of the muscle model for ideal isometric conditions

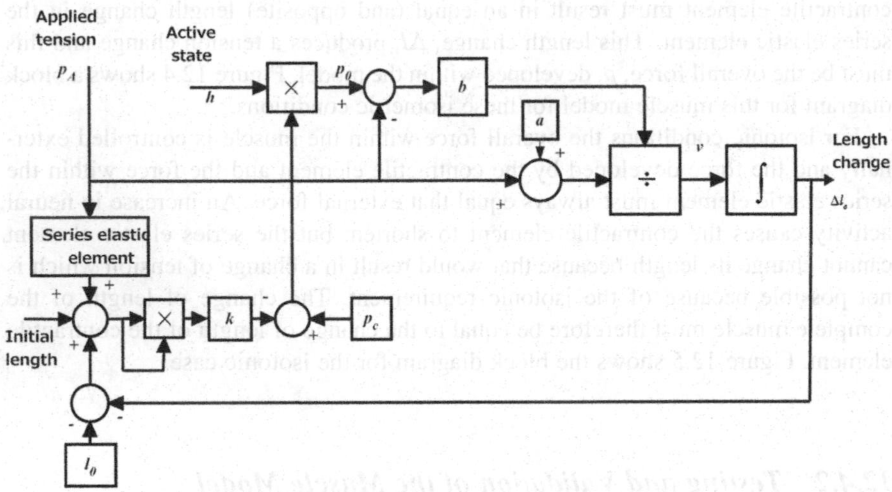

Fig. 12.5 Block diagram of the muscle model for ideal isotonic conditions

case is a combination of the block diagram for the isometric case and the block diagram for the isotonic case.

Results obtained by simulation of the muscle for the after-loaded isotonic case are shown in Fig. 12.7 for one value of applied tension (0.1515 N) along with the corresponding experimental results found by Ritchie and Wilkie [18]. The level of quantitative agreement between the experimental findings and the simulation results is encouraging during the initial phase of the twitch response (from 0 to about 0.25 s), but the simulation results show that the peak is reached at a time which is about 0.05 s after the time of the peak in the experimental response. Similar results were found for three other values of applied tension, corresponding to other experimental cases considered by Ritchie and Wilkie [18]. The overall level of agreement found in these other cases is similar to that found for the case

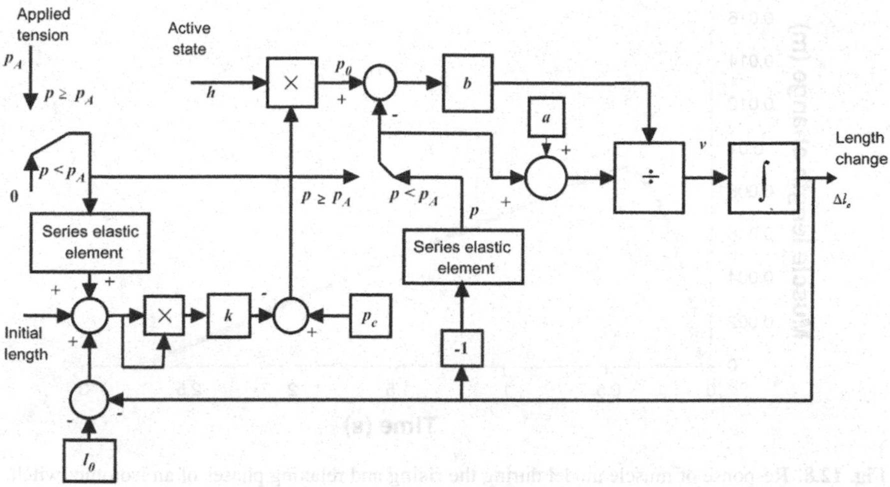

Fig. 12.6 Block diagram of the muscle model for the after-loaded isotonic case

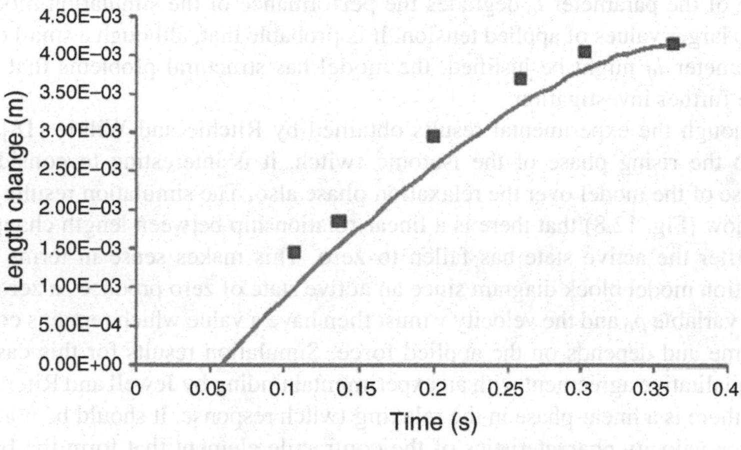

Fig. 12.7 Plot of muscle length versus obtained from the model for the after-loaded isotonic case with an applied tension of 0.1515 N (Data points shown are from Ritchie and Wilkie [18])

shown in Fig. 12.7 but, especially for smaller values of applied tension, the level of agreement between the model results and the experimental data was least satisfactory near the peak of the twitch.

Parameter sensitivity analysis was performed on the model to investigate the effect of small changes of parameters on the model variables. The findings showed that near the peak of the twitch the parameter with most influence is l_0 and that an increase in the value of this parameter of about 0.3 % is sufficient to give a significant improvement in the fit between the simulation model results and

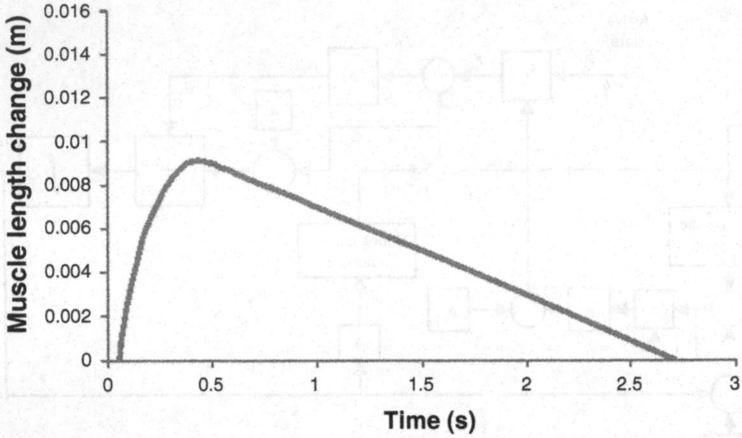

Fig. 12.8 Response of muscle model during the rising and relaxing phases of an isotonic twitch, for applied tension of 0.0436 N

experimental findings for the two smaller values of tension. However, that same change of the parameter l_0 degrades the performance of the simulation model for the two larger values of applied tension. It is probable that, although a small change of parameter l_0 might be justified, the model has structural problems that would require further investigation.

Although the experimental results obtained by Ritchie and Wilkie [18] apply only to the rising phase of the isotonic twitch, it is interesting to consider the response of the model over the relaxation phase also. The simulation results in this case show (Fig. 12.8) that there is a linear relationship between length change and time, after the active state has fallen to zero. This makes sense in terms of the simulation model block diagram since an active state of zero produces a zero value for the variable p_0 and the velocity v must then have a value which remains constant with time and depends on the applied force. Simulation results for this case also show qualitative agreement with an experimental finding by Jewell and Ritchie [19] where there is a linear phase in the relaxing twitch response. It should be noted that the force-velocity characteristics of the contractile element that form the basis of the representation of (12.2) are not strictly valid during the relaxing phase. There appears to be a lack of published experimental data for negative values of muscle velocity but the model presented here can be modified readily to accommodate other forms of force-velocity curve.

12.4.3 Applications of the Muscle Model

Although developed and tested for the after-loaded isotonic twitch experiment the model may be adapted for a range of other conditions. The most obvious are models

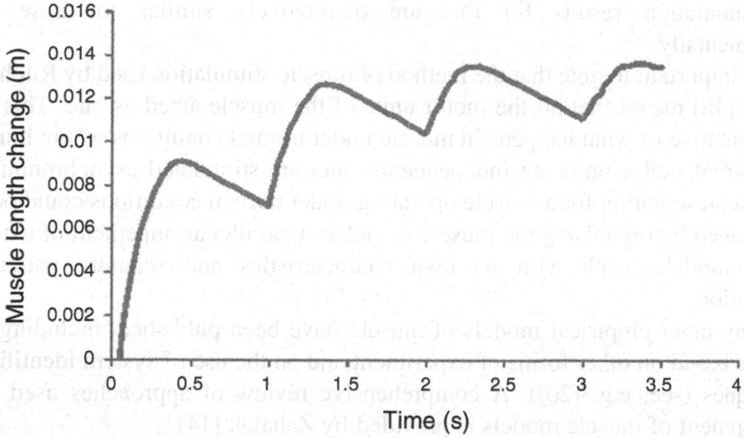

Fig. 12.9 Model response to repeated neural stimuli at frequency of 1 per second. Isotonic conditions with applied tension of 0.0436 N

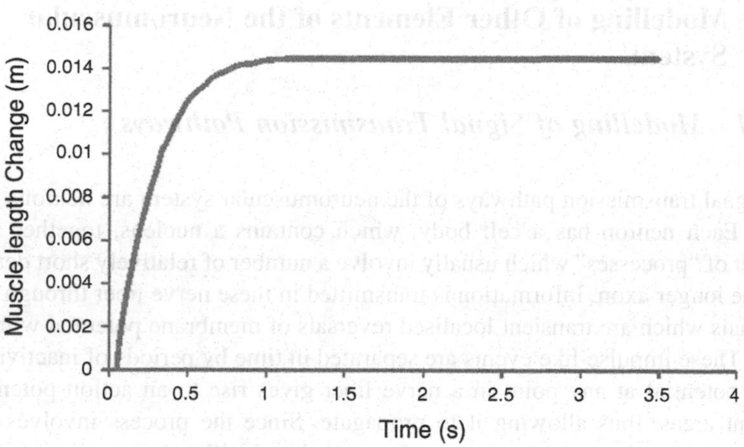

Fig. 12.10 Model response to repeated neural stimuli at frequency of 12 per second. Isotonic conditions with applied tension of 0.0436 N

for isometric and isotonic conditions which are elements within the after-loaded isotonic model. However, it is also possible to consider use of the model with a more general load. In this case the change of length of the contractile element is shared by the series elastic element and the external load and the block diagram of the model has to be changed accordingly.

The model may also be used to simulate the application of repeated neural stimuli at various frequencies for isometric or isotonic conditions. The results shown in Figs. 12.9 and 12.10 demonstrate clearly the phenomenon of twitch summation leading, in the limit, to a steady contraction and, eventually to tetanus.

The simulation results for this are qualitatively similar to those found experimentally.

It is important to note that the method of muscle stimulation used by Ritchie and Wilkie [18] meant that all the motor units of the muscle acted as one. This is not representative of what happens in muscle under normal conditions where hundreds of different motor units act independently and are stimulated asynchronously. A better representation for a muscle operating under normal conditions could perhaps be obtained by organising the muscle model as a parallel arrangement of a number of sub-models, each with its own characteristics and separate patterns of stimulation.

Many other empirical models of muscle have been published, including some that are based on other forms of experiment and on the use of system identification techniques (see, e.g. [20]). A comprehensive review of approaches used in the development of muscle models is provided by Zahalak [14].

12.5 Modelling of Other Elements of the Neuromuscular System

12.5.1 *Modelling of Signal Transmission Pathways*

The signal transmission pathways of the neuromuscular system are neurons (nerve cells). Each neuron has a cell body, which contains a nucleus, together with a number of "processes" which usually involve a number of relatively short dendrites and one longer axon. Information is transmitted in these nerve fiber through action potentials which are transient localised reversals of membrane potential within the nerve. These impulse-like events are separated in time by periods of inactivity. An action potential at any point in a nerve fiber gives rise to an action potential in adjacent areas, thus allowing it to propagate. Since the process involves active transport of ions across membranes there can be significant transmission delays, with large fiber having higher velocities of propagation. Since the duration of each action potential is normally very short compared with the time interval between action potentials, it is possible to consider these as point events. Trains of action potentials may be regarded, for most purposes, as realisations of stochastic point processes. At junctions the activity spreads into all the branches. Action potentials are commonly referred to as "spike trains" and, in for the purposes of analysis of the neuromuscular system, can be considered to carry information through a form of pulse frequency modulation.

One approach to analysis involves conversion of the inherently discrete signals within nerves into a quasi-continuous form in terms of the instantaneous frequency. This is defined as the inverse of the time interval between adjacent spikes. Spike trains can be observed at any point on the nerve fiber and, from measurements at appropriate points within a network of nerves, it is possible to define the relevant signals either in terms of the times of occurrence of the recorded spikes or through the corresponding instantaneous frequencies. Both have been shown to be useful for the purposes of system identification and for modelling.

The connecting point between one nerve cell and another is a "synapse". The effect of a nerve impulse in one fiber at a synapse may not be large enough to generate an action potential in the other nerve. The cumulative effect of many nerve impulses may be needed, especially as synaptic connections may be "facilitatory" or "inhibitory" so that synapses should be viewed as being like a summing point within an engineering block diagram, but with some additional properties such as attenuation and some nonlinear effects such as a threshold.

The investigation of the natural interactions that occur between the neurones that form the communication network of the neuromuscular system has been facilitated, in recent years, by the application of system identification methods to spike trains. One very successful approach to spike train analysis is based on point-process principles (see, e.g. [21]) using times of occurrence of spikes. Frequency-domain analysis methods have been found to be particularly helpful in this context and have led to interesting hypotheses about the generation and communication of sensory information within the system [22]. The GAP3 package, developed in the 1980s and early 1990s by the Computational and Experimental Science Group at the University of Glasgow, was one early example of a system identification package of this kind. It could be used for the estimation of time-domain and frequency-domain properties of both conventional sampled signals and point processes and was extensively used for studies involving various different elements of the neuromuscular system [23, 24]. The testing of system identification methods, such as those used in GAP3, using simple and well-understood simulation models allowed practical limitations of the identification methods to be investigated under closely-controlled and well-understood conditions. The model neurones involved simple integrate and fire descriptions with appropriate thresholds, pure time delays to represent transmission along axons and positive and negative summation at synaptic junction models to represent excitatory and inhibitory effects. Following completion of this testing process using simulation the identification tools could be used with much more confidence on real experimental data.

Simple simulation models can thus make an important contribution to the testing of system identification methods, allowing identification and digital signal processing tools to be investigated initially on well-understood systems and signals, under closely controlled conditions. The same simple simulation models are also very useful in training new users of system identification tools, to allow them to gain practical experience before attempting to work with real experimental data. An interactive tutorial package was developed at the University of Glasgow to provide an introduction to neural signal analysis techniques and the use of the associated analysis software within the GAP3 package. Known as GAPTUTOR, this package allowed data files with known properties to be created and used as input to the analysis routines in GAP3 [25]. It is clear that the inclusion of simulation facilities within tutorial packages of this kind could have similar advantages in other fields.

An interesting paper which deals with a range of fundamental issues concerning the processing of information within the nervous system and the development and testing of related models was published by Moore in 1980 [26]. As pointed out by Moore, the fitting of mathematical models to experimental input-output data is often rejected by physiologists as it does not relate to the underlying biological mechanisms. However, he suggests that approaches of this kind based on system identification methods, combined with sound principles of experiment design and physiological insight can lead to mathematical models that prove to be highly significant. In this context he particularly mentions the work of Hodgkin et al. [27] who, using a specific experimental technique (the voltage clamp), separated the currents resulting from step changes in membrane potential into two components associated with sodium and potassium. Differential equations were then fitted to the data, leading to the classic Hodgkin and Huxley description [2]. In order to establish the dependence of parameters of this empirical model on membrane potential, additional experiments were carried out. Moore claims that the success of Hodgkin and Huxley's approach can be attributed to careful experimental design and, in particular, the choice of a step function as the input for the system identification process, rather than the choice of the model structure. As mentioned in Sect. 12.2, recent research has been published relating to the quality of the Hodgkin-Huxley model and presenting new evidence of its underlying validity and range of applications [3]. Although written more than three decades ago, most of the content of Moore's paper is still highly relevant and should be of interest for anyone using modelling and simulation methods in biology.

12.5.2 Modelling of Sensory Receptors

The most important sensory receptors within the neuromuscular control system are muscle spindle receptors and Golgi tendon organs (see, e.g. [10]). In functional terms, muscle spindle receptors lie in parallel with the extrafusal fiber of the load-bearing muscle (as shown in Fig. 12.1) and may be regarded, in a simplified way, as being a form of transducer which detects stretching of the extrafusal muscle but can also be influenced by signals from the nervous system. It can therefore be thought of, in engineering terms, as being rather like a strain gauge that has a gauge factor that is under some form of external control. Muscle spindles consist of a small number of highly specialised muscle fiber (the "intrafusal" fiber) which are arranged in parallel and are contained partially within a capsule of connective tissue which is filled with fluid (see, e.g. Boyd [28]). The intrafusal fiber are much shorter than the extrafusal fiber. Most skeletal muscles contain a number of muscle spindle receptors, operating apparently in a parallel fashion. In many muscles more axons are associated with muscle spindles than with the extrafusal muscle fiber of the load-bearing muscle and this emphasises the importance of these receptors within the complete neuromuscular system.

A single muscle spindle may be regarded as a sensor which responds to changes of length of the muscle in which it is embedded and responds also to neural inputs which cause local contractions within the intrafusal fiber. At least three different types of intrafusal fiber have been identified. These are the "nuclear-bag" fiber, of which two distinct types have been found and are referred to as "bag_1" and "bag_2" fiber [28]. These are distinguished by the fact that bag_1 fiber are known to be more velocity sensitive than the other types of intrafusal fiber. The third group involves shorter fiber and these are known as "nuclear-chain" fiber. The bag_2 and chain fiber are known to contribute mostly to length sensitivity within the overall muscle spindle response. Most muscle spindles have one of each type of nuclear-bag fiber together with a number of nuclear-chain fiber.

The outputs of the muscle spindle are transmitted to the spinal cord as spike train signals carried by two types of sensory neurone, termed "primary" (or "Group Ia") and "secondary" (or "Group II") axons. The frequency of these pulse trains depends, in part, on muscle length and can be modulated by changes of muscle length. The primary afferent endings are located in the equatorial parts of the three types of intrafusal fiber and it is from these endings that the information about length and velocity is transmitted to the spinal cord and thus to the rest of the nervous system. The secondary afferent axon has endings on both the bag_2 and chain fiber but these are mainly located away from the equatorial region and are primarily length sensitive.

As well as the sensory axons, the intrafusal fiber are innervated (as mentioned above) by the axons of other cells lying within the spinal cord. These are known as the "gamma" motoneurones or "fusimotor" neurones and they are much smaller in diameter than the alpha-motoneurones of the main muscle. They are categorised as "gamma-dynamic" and "gamma-static" axons depending on their effects (see, e.g. [28–30] for details). Activity in these neurones alters the response of the Ia and II axons to imposed length changes. The gamma-dynamic fusimotor neurones innervate bag_1 fiber (also known as the "dynamic nuclear bag" fiber) while the gamma-static fusimotor neurones may innervate either the nuclear-chain fiber or the bag_2 fiber (also known as the "static nuclear bag" fiber), or both of these.

Experimentation presents considerable difficulties and measurement of variables within the muscle spindle, such as the tension within a specific intrafusal fiber, is possible only in experiments on isolated spindles. The only output variables that are readily available are the spike train signals in the Group Ia and Group II axons.

In dynamic terms the responses of muscle spindle receptors to time-varying length changes of the parent muscle are different for the primary and secondary sensory neurones. If the instantaneous frequency of action potentials is used as a measure of activity in the sensory neurones it becomes clear, from examination of measured data, that the primary response is highly sensitive to muscle velocity as well as to muscle length, while the secondary response shows a much lower level of velocity sensitivity. The fusimotor axons, which innervate the intrafusal muscle fiber, are known to modify the muscle spindle primary and secondary dynamic responses to imposed length changes. This rate sensitivity of muscle spindle primary endings is believed by many to have a damping and stabilising role within the neuromuscular control system, in much the same way that velocity feedback is designed into many engineering control systems.

Clearly, from its structure, the muscle spindle should be viewed as a multi-input, multi-output system. It involves an unusual combination of continuous and discrete variables. Instantaneous frequency measures of nervous signal data have been applied in system identification studies of the muscle spindle (e.g. [31]) and also the Golgi tendon organ (e.g., [32]), but this approach is appropriate only for situations involving periodic inputs of low frequency and rates of change of input that are small. The problem can be seen most clearly from the periodic case where, at high frequencies of input, the number of spikes occurring within one period of the stimulus may be too small to provide a useful measure of system behaviour. In the real system it is clear that averaging takes place over the outputs from many different receptors but normally measurements of afferent activity are made from an electrode on a single axon.

Research on the analysis of point processes suggests that random signal testing applied to both continuous and discrete inputs may be particularly useful in gaining a better understanding of the interactions that take place within the system [33]. Frequency-domain analysis methods have been found to provide useful insight [23].

Although the muscle spindle was often described in the past in terms of a single-input single output type of representation, with muscle length as an input and the Ia or II afferent axon signals as output, it is now recognised as a multi-input multi-output system. This may be represented, in simplified form, by the block diagram of Fig. 12.11. In this diagram the first main block represents mechanical properties of the intrafusal fiber which have mechanical and neural inputs. The second block represents the processes involved in the conversion of strain within specific regions of the intrafusal fiber into receptor potentials within the surrounding nerve endings, together with encoder elements which combine the receptor potentials to generate pulse train outputs in the Ia and II afferent axons.

Some of the research on modelling of muscle spindles reported in the literature has involved the application of system identification and parameter estimation methods using experimental data. Other work has involved the use of relatively simple visco-elastic models based on the simplified models representing the mechanical properties of the intrafusal fiber, usually in lumped parameter form. Many early models of the muscle spindle, although basically single-input single-output descriptions, attempted to relate properties of identified transfer function descriptions found from experimental testing to the mechanical properties of intrafusal fiber (see e.g. [34–36]). Experimental modelling techniques of this kind also allow investigation of the effect of fusimotor inputs on transfer functions relating mechanical inputs to afferent signal outputs by attempting to find changes in the estimated model parameters for different conditions of fusimotor stimulation (see, e.g. [31]).

The encoder elements within the second main block of Fig. 12.11 attracted less attention in early models of the muscle spindle because less was known about the mechanisms responsible for the summation and interaction of receptor potentials and the processes involved in the generation of the output pulses. Early models also

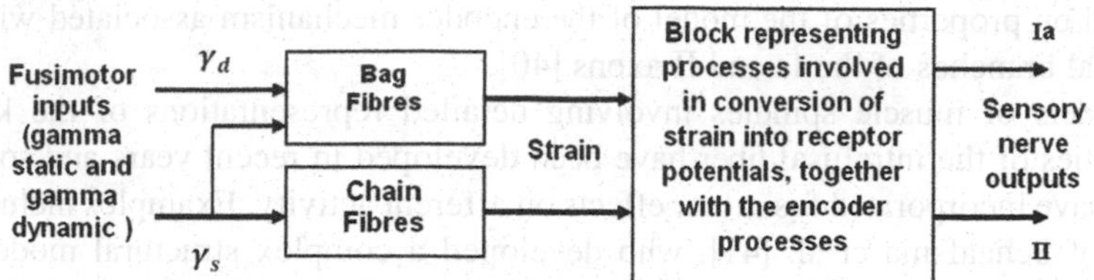

Fig. 12.11 Block diagram of muscle spindle showing fusimotor inputs. It should be noted that the nuclear bag and nuclear chain fibers are also subjected to mechanical stimuli through changes of length of the extrafusal muscle

involved relatively simple modelling of the effects of fusimotor stimulation (see, e.g. [37]) since there was, again, much uncertainty about the physiological mechanisms involved.

In some cases the modelling of intrafusal fibers has been based on the structure of models of extrafusal muscle, with parameter values suitably modified on the basis of available anatomical and physiological knowledge. It is interesting to note that in 1983 a comment was made by Murray-Smith and Rosenberg [38] that relatively little use had been made of dynamic modelling and computer simulation methods in muscle spindle research over the previous decade. A plea was made for more use of these methods in addressing problems that were viewed as important at that time, many of which were associated with areas of uncertainty and incomplete knowledge. One topic that was highlighted concerned the mechanisms responsible for changes of responsiveness of Ia or II axons to muscle stretching in the presence of fusimotor inputs. Questions were also raised about details of the Ia and II terminal branches and how stretching of the sensory endings in the different intrafusal fiber leads to the afferent signals observed experimentally in the Ia and II axons. It was suggested that computer simulation techniques should be used to test any suggested interpretation of experimental data. This could involve the testing of different model structures using analysis techniques which are the same as those applied to measured response data from the real system.

One early example of an approach of this kind involved investigation of nonlinear muscle spindle behaviour in experiments involving sinusoidal length changes applied to the main muscle [39]. It is well known that tests of this kind may lead to an afferent discharge which is phase-locked or synchronised with the input sinusoid. It is also known that experiments involving the application of regularly repeating patterns of fusimotor activity to a muscle spindle may show fixed and exact relationships in the pulse frequencies of the gamma-efferent and afferent signals, in a phenomenon which is termed "driving". This driving effect appears to over-ride any other sensory information that might otherwise be expected in the afferent signal. Although a number of theories had been suggested concerning the origins of the driving and phase-locking effects, simulation results suggest that the synchronising and phase-locking phenomena observed with periodic length signals and fusimotor input patterns could be similar in origin. They depend on properties of the model of the encoder mechanism associated with the terminal branches of the Ia and II axons [40].

Models of muscle spindles involving detailed representations of the known properties of the intrafusal fiber have been developed in recent years and some of these have incorporated fusimotor effects on afferent activity. Examples include the work of Schaafsma et al. [41], who developed a complex structural model that involved fusimotor effects on the primary discharge. Their model included two sub-models representing the different types of intrafusal fiber. A further development was a black-box type of model by Maltenfort and Burke [42] which included the effects of static and dynamic fusimotor activation, but was found to produce predictions of peak primary firing in the presence of dynamic fusimotor inputs that were unrealistically high. This model also had a number of other features that were judged to be unrealistic [42]. Lin and Crago [43] developed a combined model of extrafusal muscle and a muscle spindle receptor. In this case a Hill type model was used to represent the extrafusal muscle fiber and the spindle model was structural in form, with three different types of fiber represented. The primary afferent response predictions of this model by Lin and Crago were judged to be realistic for ramp-and-hold and sinusoidal stretch inputs, especially for cases involving fusimotor inputs. As in the muscle model discussed in Sect. 12.4.1, the intrafusal muscle force is determined by the product of the activation level and the force found from the force-length and force-velocity characteristics. It should be noted that this type of approach to the modelling of intrafusal fiber was also a feature of at least one much earlier model of the muscle spindle [44].

In 2006 a major step forward was made with the publication of a highly detailed model of the muscle spindle by Mileusnic et al. [45] which incorporated three nonlinear intrafusal fiber sub-models representing bag_1, bag_2 and chain fiber. A single set of model parameters was established and optimised for a number of experimental data sets involving a variety of test input types. Model validation work involved data sets that were different from those used in establishing the model parameters during the initial development. It is of interest to note that the papers by Mileusinic et al. [45] and by Lin and Crago [43] both highlight areas that were discussed in the critical review of muscle spindle models by Murray-Smith and Rosenberg in 1983 [38]. One point of discussion in that paper related to the fact that the origin of changes of responsiveness of Ia or II axons to muscle stretching in the presence of fusimotor inputs were far from clear. The possible alteration of the mechanical properties of intrafusal fiber by fusimotor inputs could well account for observed changes in Ia and II responses to imposed length changes but there were also uncertainties about the region where different terminal branches came together to form the sensory axons leaving the muscle spindle. It was possible that fusimotor activity might switch the site where pulse activity is initiated within the Ia and II terminal branches. These issues are addressed in some detail in the model of Mileusnic et al. [45] in the light of experimental evidence gathered in recent years and it is fair to say that this model is more comprehensive and more fully tested than previous models of the muscle spindle and provides a concise summary of knowledge about the muscle spindle at its date of publication.

Golgi tendon organs lie in series with the load-bearing muscle and give neural output signals which are believed to be proportional to the tension within the main muscle with which they are associated. It is important to note that Golgi tendon organs are simpler than muscle spindles and, unlike muscle spindle receptors, the neural output of a Golgi tendon organ is not affected by any input from the nervous system. They therefore have no equivalent of the fusimotor innervations of muscle spindles and are essentially passive sense organs. A paper by Mileusnic et al. [46] is a fairly recent contribution describing models of the Golgi tendon organ and this, together with earlier published work on the modelling of this type of sensory receptor, is of potential interest in terms of models of the complete neuromuscular control system.

Both muscle spindle and Golgi tendon organ receptors are believed to be involved in the control mechanisms for maintaining posture and initiating movement. Models of muscle spindles and tendon organs, together with models of active skeletal muscle, have proved useful in testing hypotheses about these control mechanisms. Issues, such as the role of tendon organs which provide positive force feedback during locomotion, have been addressed successfully using models. Also, the significance of the nonlinear properties of muscle in ensuring that the gain of positive force feedback loops fall to levels that ensure stability as muscles shorten would not have become clear without the use of modelling and simulation methods (see, e.g. [4, 9, 10]).

12.5.3 The Testing of Models of Sensory Receptors

In investigating the role of muscle spindle receptors within the systems for the control of movement there are some fundamental problems of model validation. Most experimental investigations relating to muscle spindles have involved isolated spindles and experimentalists such as Boyd (see, e.g. [28]), Barker et al. (see, e.g. [47]) Laporte et al. (see, e.g. [48]) and Bessou et al. (see, e.g. [49]) have been responsible for very elegant techniques which have provided valuable information about typical mechanical parameters of the different types of intrafusal fiber. Other investigators have provided important information about receptor potentials and encoder properties. However, in terms of experimental data that could be used for external validation of a dynamic model of a single isolated muscle spindle, there are still significant difficulties. Some of these relate to the fact that experiments on isolated muscle spindles are very complex and time consuming and the properties of the isolated spindle tend to change with time over the period of the experiment. Also, it has to be accepted that the parameters of a given model are likely to be obtained from analysis of results from many different experiments on different muscle spindles. This makes it impossible to have data sets that can be used for quantitative validation purposes since the model does not represent a single muscle spindle. The best we can hope for is to have estimates of parameters, such as stiffness and viscosity in approximate lumped parameter representations. However, these data sets will probably have been gathered in a number of different experimental investigations involving different muscle spindles. As a result most physiological research involving models of muscle spindles has been concerned with qualitative rather than quantitative testing of model fidelity. A model may be accepted if it makes predictions that are broadly consistent with observations of the corresponding experimental situation.

Another limiting factor in terms of testing models of the muscle spindle is the fact that it is very difficult to obtain simultaneous records of the afferent and fusimotor activity in situations in which the muscle spindles are operating in a natural way. This is clearly very important when muscle spindle models are being developed as sub-models within a more complex representation intended to provide insight about the neuromuscular system as a whole, or to allow the testing of hypotheses about neuromuscular control. Often, relatively simple models of the elements of the system, including the muscle spindle, have proved more useful in this context than highly detailed, and thus computationally more intensive, models.

Additional information about sensory receptors, along with background information relating to other elements of the neuromuscular system, may be found at the website associated with this book. Additional details relating to the model of skeletal muscle presented in Sect. 12.4.1 are included, together with information about a linearised version of that model. From that linearised description, a model has been developed to represent the dynamic properties of intrafusal muscle fiber, leading to a simple model of a muscle spindle receptor. The website includes an outline of a face validation type of process that has been used to evaluate this muscle spindle model.

12.6 Models of the Neuromuscular Control System and the Role of Simulation in the Testing of Hypotheses

As has already been noted, one of the main objectives in developing computer-based models of skeletal muscle and of receptors is to try to gain a better understanding of the workings of the neuromuscular control system. In the case of a single muscle, the contraction of the extrafusal muscle fiber is caused by the signals transmitted to the muscle end-plates by alpha motoneurones. The activity within a single alpha motoneurone is, however, determined by many different feedback pathways, involving many different receptors, including the muscle spindle receptors, Golgi tendon organs and other receptors such as joint receptors and skin receptors which have not been mentioned previously in this account of the system.

Models of the stretch reflex system typically include a model of active skeletal muscle, models of muscle receptors and models of the relevant signal transmission pathways. An example of a recent study of this kind may be found in the work of He et al. [50] who have developed a comprehensive model which includes the dynamic properties of muscle, muscle spindles and Golgi tendon organs, feedback pathways from these sensory receptors and also incorporates a local feedback loop at the spinal cord (involving "Renshaw cells"). A useful review of competing ideas associated with reflex and voluntary components of neuromuscular control may be found in a paper by Prochazka et al. [51]. Further model-based approaches may be found in the work of Perreault et al. who have used identification-based methods within a non-parametric modelling approach [52].

Testing of a model of a very complex system of this kind raises many issues because of the uncertainties associated with the physiological experimentation. The complete model may involve elements that have been tested using data from many different experiments which may have been carried out at different times and by different investigators using approaches that differ in detail. In most cases the measurements are difficult to obtain and often information about the system under test is incomplete and sometimes very limited. In addition, the biological variability tends to be large. Validation in the sense of getting a close match between experimental results and simulation output is therefore very difficult and model validation in this field is most often a process based on expert opinion. The views of physiologists who are experienced in the experimental investigation of neuromuscular systems are central to the process and validation usually involves qualitative analysis of simulated experiments not considered in the development of the model. In some cases quantitative comparisons may also be appropriate but, usually, comparisons of trends in results as experimental conditions are varied are seen as more important than simple numerical comparisons of measured and modelled quantities for one specific operation condition.

The problems of model testing and validation are greatest when the complete neuromuscular system is considered and this is especially important since it is now possible to record from muscle spindle afferents during natural movements such as walking (see e.g. [4]). Many suggestions have been made over the years about the way in which the neuromuscular system functions, including early hypotheses such as those by Hammond et al. [53] and Marsden et al. [54] which put involved a theory based on servomechanism principles. However, it is now generally accepted that simple servomechanism-based models of this type are not suitable. The tasks performed by the neuromuscular control system are inconsistent with simple models of this kind and there is evidence that the synaptic transmission of signals from the muscle spindle and Golgi tendon organ proprioceptors is modulated by the central nervous system in a way that depends on the task being undertaken.

It is now generally believed that controlled variables within the neuromuscular system cannot be defined in a simple way because of the complex interactions that take place within muscle spindles. Experimental research results have also made it clear that positive feedback has an important role in the control of movement and Prochazka et al. [55] have provided a very interesting analysis of the implications of positive feedback in animal motor control systems using detailed block-diagram models. Recent developments have involved hypothesis testing based on computer simulation models along with careful analysis of the precise timing of events in multi-channel experimental records. From this evidence (see, e.g. Prochazka et al. [9, 56]), it has been suggested that afferent signals from the sensory receptors may have a role in decision making at the upper levels of the hierarchical control system.

The problem of understanding how stretch reflex control ends in one set of muscles at a joint and is started in a second, with a smooth transition, has been seen as particularly important for cyclic tasks such as walking. Switching actions involving some form of conditional logic are believed to be involved (see, e.g. [57]). These ideas on switching have been linked to the suggestion by Prochazka [58] that the concepts of fuzzy logic control may provide a more "biologically compatible" form of model than could be offered by other forms of logic-based description. Thus, experimental investigations and computer simulation are being brought together to allow new hypotheses to be developed and tested in an integrated way, with model testing and validation being properly recognised as an essential element in the whole process.

Another topic, which is of growing importance, concerns prosthetics. A good example is provided by the work of Markowitz et al. [59] on control systems for powered ankle-foot prostheses. The aim of the research was to make the prosthesis adaptive for variations of walking speed and evidence that reflexes contribute to changes in ankle joint dynamics in normal subjects has been used to allow development of a reflex-based controller involving a neuromuscular model. Although it does not necessarily reflect the biological control system, this controller was found to reproduce, quite closely, the muscle dynamics that would be observed from biological data [59].

A further type of application which is becoming increasingly important involves functional neuromuscular stimulation (FNS), which is concerned with rehabilitation and involves the application of controlled electrical stimulation to activate paralysed muscles for restoring postural control and movement. A control system is needed to establish which muscles should be stimulated and the patterns of stimulation necessary for given tasks. Inevitably, the design of such systems requires a good model of the system being controlled (see, e.g. [58]). Useful accounts of research on the design of closed-loop control systems involving the use of FNS techniques may be found in the work of Matjacić et al. (see, e.g. [60]). An interesting approach which involves the use of biologically inspired models in implementing control techniques for motor planning assistance is described in a paper by Conforto et al. [61]. A further development in neuroprosthetics where modelling and simulation have an important part to play relates to implantable peripheral neural interfaces. Different types of electrode have been developed to provide selective neural stimulation and modelling studies have provided information that is important for electrode shape design [62].

Quite separately from research on rehabilitation, the modelling of movement disorders is starting to attract attention, as shown by recent research by Mugge et al. [63]. A nonlinear neuromuscular model of the wrist, involving muscle dynamics and neural elements, has been developed and used to test hypotheses relating to a condition termed "dystonia". Symptoms of dystonia include repetitive movements or abnormal postures and are associated with involuntary and sustained contraction of specific muscles. Modelling and simulation results suggest that an imbalance in terms of reflex system sensitivities, combined with an element of unstable force feedback, could be shown to resemble, closely, all the features of dystonia [63].

Research on movement and multi-joint posture control in humans is a topic which is now receiving a lot of attention. An interesting account of the problems of movement control has been provided by Popović [64] and some specific movements, such as those involved in an arm reach task, have been modelled in detail (see e.g. Lan et al. [65]). Neuromuscular models are now contributing to some areas of research in the field of sports science and an interesting account of work in that field has been provided by Thaller et al. [66] who have presented a model-based formulation of ideas that are claimed to show benefits when used in the training of athletes. Hill's equation ((12.1) in Sect. 12.4) provides the basis of that work and methods are proposed to allow relevant parameters to be estimated for individual human subjects which can then be used to develop enhanced training strategies.

Research on human movement control has obvious links with robotics and biologically-inspired control schemes are now seen to have possible benefits in the development of legged robots. The introduction of robotic systems based on biomimetic principles depend on our growing understanding of sensory receptors and neuromuscular systems. Equally, this research on robotics is generating new ways of thinking about biological control concepts within the neuromuscular system. Prochazka [67] has suggested that conclusions reached by Nelson et al. [68, 69] concerning sensory feedback in biomimetic robots are consistent with our current understanding of the biological system and that, using concepts from fuzzy control, we can largely explain the mechanisms of balance and stable locomotion. Clearly, this is a very interesting area of science where new interactions between neurophysiologists, biomedical engineers and those involved in the design of robotic systems are likely to produce interesting developments in future.

12.6.1 Testing of Models of the Neuromuscular Control System

Methods of external validation that are based on precise quantitative measures of model performance are probably of less relevance in this area of physiology than in engineering applications, or in some other areas of science, due to the highly parallel and redundant internal structure of the system and the problems of obtaining measurements of internal signals from the intact system under normal conditions of operation. The use of face validation in the assessment of the model behaviour, in the context of the intended application, is perhaps more important than traditional quantitative methods. In such situations experienced experimental neurophysiologists, working together with simulation, modelling and control systems specialists who have the necessary physiological understanding, can certainly contribute more to the advancement of research in this field than any one of those groups when working in isolation.

One important observation is that some research groups working on problems of neuromuscular control make a point of making public the full details of simulation models that they have developed and applied. This has been an important feature of publications from the research group at the University of Alberta in Canada for some considerable time. Details of simulation models used in their work have been included in many of their papers, with detailed information about model structure, parameter values and Simulink diagrams (e.g. [55, 58, 70]). This is extremely helpful as it allows other researchers to repeat the simulation work reported in these papers and should assist the process of model assessment and acceptance. It is interesting to note that members of other research groups in the field of neuromuscular systems are now also publishing full details of simulation models used in the course of their investigations (e.g. [45, 65]).

At the level of models of human movement, one interesting development in terms of model validation has been reported by Abraham et al. [71], who used a Turing-like test to investigate the relative merits of three different models of handshake movements. The test was administered through a tele-robotic system and task involved a human subject trying to assess which model produced a handshake most like that of a human. The results suggested improvements that could be made in terms of models for this system.

It is clear from this case study that many interesting issues arise in investigating elements of the neuromuscular system. Modelling and computer simulation methods have been applied with some success in this field and it is clear that questions of model quality and the value of a particular model for a specific intended application are of key importance.

References

1. Murray-Smith DJ, Zhao S (2007) A graduate-level multi-purpose educational case study on modelling and simulation of the neuromuscular system. In: Zupancic B, Karba R, Blazic S (eds) Proceedings EUROSIM 2007, 9–13 September 2007, Ljubljana, Slovenia. SLOSIM, Ljubljana
2. Hodgkin AL, Huxley AF (1952) A quantitative description of membrane current and its application to conduction and excitation in nerve. J Physiol 117(4):500–544
3. Lindsay KA, Rosenberg JR, Tucker G (2004) A note on the discrepancy between the predicted and observed speed of the propagated action potential in the squid giant axon. J Theor Biol 230 (1):39–48
4. Prochazka A, Gillard D (1997) Sensory control of locomotion. In: Proceedings of the American control conference, Albuquerque, New Mexico, June 4–6 1997. American Automatic Control Council, Evanston, pp 2846–2850
5. Zajac FE (1989) Muscle and tendon: properties, models, scaling and applications to biomechanics and motor control. CRC Crit Rev Biomed Eng 17:359–411

6. Binder M, Heckman C (1996) The physiological control of motoneuron activity. In: Rowell L, Shepherd J (eds) Handbook of physiology, exercise: regulation and integration of multiple systems. Oxford University Press, New York, pp 3–53
7. Loeb GE, Brown IE, Cheng EJ (1999) A hierarchical foundation for models of sensorimotor control. Exp Brain Res 126:1–18
8. Valero-Cuevas FJ, Hoffmann H, Kurse MU et al (2009) Computational models for neuromuscular function. IEEE Rev Biomed Eng 2:110–135
9. Prochazka A, Gritsenko V, Yakovenko S (2002) Sensory control of locomotion: reflexes versus higher-level control. In: Gadevia SG, Proske U, Stuart DG (eds) Sensorimotor control. Kluwer Academic/Plenum, London
10. Prochazka A, Ellaway P (2012) Sensory systems in control of movement. Compr Physiol (Am Physiol Soc) 2(4):2615–2627
11. Keifer J, Houk JC (1994) Motor functions in the cerebellorubrospinal system. Physiol Rev 74 (3):509–542
12. Hatze H (1997) A myocybernetic control model of skeletal muscle. Biol Cybern 25:103–119
13. Hatze H (1978) A general myocybernetic control model of skeletal muscle. Biol Cybern 28:143–157
14. Zahalak GI (1992) An overview of muscle modelling. In: Stein RB, Peckham PH, Povović DP (eds) Neural prostheses. Replacing motor function after disease or disability. Oxford University Press, Oxford, pp 17–57
15. Murray-Smith DJ (1994) The development, validation and application of a simulation model of active skeletal muscle. In: Halin J, Karplus W, Rimane R (eds) First joint conference of international simulation societies proceedings, August 22–25, 1994, ETH Zurich, Switzerland. The Society for Computer Simulation International, San Diego, pp 600–604
16. Hill AV (1922) The maximum work and mechanical efficiency of human muscles and their most economical speed. J Physiol 56:19–41
17. Hill AV (1938) The heat of shortening and dynamic constants of muscle. Proc R Soc Lond B126:136–195
18. Ritchie JM, Wilkie DR (1958) Dynamics of muscular contraction. J Physiol 143:104–113
19. Jewell BR, Wilkie DR (1960) The mechanical properties of relaxing muscle. J Physiol 152:30–47
20. Gollee H, Murray-Smith DJ, Jarvis JC (2001) A nonlinear approach to modelling of electrically stimulated skeletal muscle. IEEE Trans Biomed Eng 48(4):406–415
21. Brillinger DR, Lindsay KA, Rosenberg JR (2009) Combining frequency and time domain approaches to systems with multiple spike train input and output. Biol Cybern 100(6):459–474
22. Rosenberg JR, Amjad AM, Breeze P et al (1989) The Fourier approach to the identification of functional coupling between neuronal spike trains. Prog Biophys Mol Biol 53:1–31
23. Halliday DM, Murray-Smith DJ, Rosenberg JR (1992) A frequency domain identification approach to the study of neuromuscular systems – a combined experimental and modelling approach. Trans Inst Meas Control 14:79–90
24. Halliday DM, Rosenberg JR, Amjad AM et al (1995) A framework for the analysis of time series/point process data – theory and application to the study of physiological tremor, single motor unit discharges and electromyograms. Prog Biophys Mol Biol 64(2/3):237–278
25. Murray-Smith DJ, Murray-Smith E, Rosenberg JR et al (1995) GAPTUTOR – a simulation-based tutorial introduction to methods for the analysis of neuronal interactions. In: Breitenecker F, Husinsky I (eds) Proceedings of EUROSIM '95 conference, Vienna, February 1995. Elsevier, Amsterdam
26. Moore GP (1980) Mathematical techniques for studying information processing by the nervous system. In: Pinsker HM, Willis WD Jr (eds) Information processing in the nervous system. Raven Press, New York, pp 17–30
27. Hodgkin AL, Huxley AF, Katz B (1952) Measurement of current-voltage relations in the membrane of the giant axon of *Loligo*. J Physiol 116:424–448

28. Boyd IA (1976) The response of fast and slow nuclear bag fiber and nuclear chain fiber in isolated cat muscle spindles to fusimotor stimulation, and the effect of intrafusal contraction on the sensory endings. Q J Exp Physiol Cogn Med Sci 61(3):203–254
29. Crowe A, Matthews PBC (1964) The effects of stimulation of static and dynamic fusimotor fiber on the response to stretching of the primary endings of muscle spindles. J Physiol 174:109–131
30. Emonet-Denand F, Laporte Y, Matthews PBC et al (1977) On the subdivision of static and dynamic fusimotor axons on the primary endings of the cat muscle spindle. J Physiol 268:835–850
31. Maclaine CG, McWilliams PN, Murray-Smith DJ et al (1977) A possible mode of action of fusimotor axons as revealed by system identification techniques. Brian Res 135:351–357
32. Houk JC, Simon W (1967) Responses of Golgi tendon organs to forces applied to the muscle tendon. J Neurophysiol 30:1466–1481
33. Rosenberg JR, Murray-Smith DJ, Rigas A (1982) An introduction to the application of system identification techniques to elements of the neuromuscular system. Trans Inst Meas Control 4 (4):187–201
34. Poppele RE, Bowman RJ (1970) Quantitative description of linear behavior of mammalian muscle spindles. J Neurophysiol 33:59–72
35. Rudjord T (1970) A second order mechanical model of muscle spindle primary endings. Kybernetik 6:205–213
36. Hasan Z (1983) A model of spindle afferent response to muscle stretch. J Neurophysiol 49:989–1006
37. Chen W, Poppele R (1978) Small-signal analysis of response of mammalian muscle spindles with fusimotor stimulation and a comparison with large-signal responses. J Neurophysiol 41:15–27
38. Murray-Smith DJ, Rosenberg JR (1983) Models of the muscle spindle: a case study in physiological system simulation. In: Proceedings 1st European Simulation Congress, 12–16 Sep 1983, Aachen, Germany, Informatik Fachberichte 71. Springer, Berlin, pp 519–523
39. Hulliger H, Matthews PBC, Noth J (1977) Static and dynamic fusimotor action on the response of Ia fiber to low frequency sinusoidal stretching of widely ranging amplitude. J Physiol 267:811–838
40. Dutia MB, Murray-Smith DJ, Rosenberg JR et al (1977) The dependence of 'driving' of Ia axons on muscle length and fusimotor sitmulation frequency. J Physiol 273(2):30P–31P
41. Schaafsma A, Otten E, Van Willigen JD (1991) A muscle spindle model for primary afferent firing based on a simulation of intrafusal mechanical events. J Neurophysiol 65:1297–1312
42. Maltenfort MG, Burke RE (2003) Spindle model responsiveness to mixed fusimotor inputs and testable predictions of beta feedback effects. J Neurophysiol 89(5):2797–2809
43. Lin CC, Crago PE (2002) Structural model of the muscle spindle. Ann Biomed Eng 30 (1):68–83
44. Murray-Smith DJ (1970) An application of modelling techniques to the neuromuscular control system, PhD dissertation, University of Glasgow, UK
45. Mileusnic MP, Brown IE, Lan et al (2006) Mathematical models of proprioceptors 1. Control and transduction in the muscle spindle. J Neurophysiol 96(4):1772–1788
46. Mileusnic MP, Loeb GE (2006) Mathematical models of proprioceptors. II. Structure and function of the Golgi tendon organ. J Neurophysiol 96:1789–1802
47. Barker D, Stacey MJ, Adal MN (1970) Fusimotor innervations in the cat. Philos Trans R Soc Lond B 258(825):315–346
48. Laporte Y, Emonet-Dénand F (1976) The skeleto-fusimotor innervation of cat muscle spindle. Prog Brain Res Underst Stretch Reflex 44:99–105
49. Bessou P, Pages B (1975) Cinematographic analysis of contractile events produced in intrafusal muscle fiber by stimulation of static and dynamic fusimotor axons. J Physiol 252:397–427

50. He J, Maltenford MG, Wang Q et al (2001) Learning from biological systems: modelling neural control. IEEE Control Syst Mag 21:55–69
51. Prochazka A, Clarac F, Loeb GE et al (2000) What do *reflex* and *voluntary* mean? Modern views on an ancient debate. Exp Brain Res 130:417–432
52. Perreault EJ, Crago PE, Kirsch RF (2000) Estimation of intrinsic and reflex contributions to muscle dynamics: a modelling study. IEEE Trans Biomed Eng 47(11):1413–1421
53. Hammond PH, Merton PA, Sutton GG (1966) Nervous gradation of muscular contraction. Br Med Bull 12:214–218
54. Marsden CD, Merton PA, Morton HB (1977) The sensory mechanism of servo actionin human muscle. J Physiol 265(2):523–535
55. Prochazka A, Gillard D, Bennett DJ (1997) Implications of positive feedback on the control of movement. J Neurophysiol 77:3237–3251
56. Yakovenko S, Chetsenko V, Prochazka A (2004) Contribution of stretch reflexes to locomotor control: a modelling study. Biol Cybern 90(2):146–155
57. Prochazka A (1999) Quantifying proprioception, In: Binder MD (ed) Progress in Brain Research, 123, Chapter 11, Elsevier, Amsterdam, pp 133–142
58. Prochazka A (1996) The fuzzy logic of visuomotor control. Can J Physiol Pharmacol 74:456–462
59. Markowitz J, Krishnaswarmy P, Eilenberg MF et al (2011) Speed adaptation in a powered transtibial prosthesis controlled with a neuromuscular model. Philos Trans R Soc B 366:1621–1631
60. Matjačić Z, Hunt J, Gollee H, Sinkjaer T (2003) Control of posture with FES systems. Med Eng Phys 25(1):51–62
61. Conforto S, Bernabucci I, Severini G et al (2009) Biologically inspired modelling for the control of upper limb movements: from concept studies to future applications. Front Neurosci 3, Article 3. Published online doi:10.3389/neuro.12.003.2009
62. Raspopovic S, Capogrosso M, Badia J (2012) Experimental validation of a hybrid computational model for selective stimulation using transverse intrafascicular multichannel electrodes. IEEE Trans Neural Syst Rehabil Eng 20(3):395–404
63. Mugge W, Munts AG, Schouten AC et al (2012) Modeling movement disorders – CRPS-related dystonia explained by abnormal proprioceptive reflexes. J Biomech 45:90–98
64. Popović D (2000) Control of movements. In: Bronzino JD (ed) The Biomedical Engineering Handbook, 2nd edn, vol II, CRC Press, Boca Raton, FL, pp 165.1–165.20
65. Lan N, Song D, Mileusnic M et al (2005) Modeling spinal sensorimotor control for reach task. In: Proceedings of 2005 I.E. Engineering in Medicine and Biology 27th annual conference, Shanghai, China, September 1–4 2005, pp 4404–4407
66. Thaller S, Tilp M, Sust M (2010) The effect of individual neuromuscular properties on performance in sports. Math Comput Model Dyn Syst 16(7):417–430
67. Prochazka A (2002) The man-machine analogy in robotics and neurophysiology. J Autom Control Univ Belgrade 12(1):4–8
68. Nelson GM, Quinn RD (1999) Posture control of a cockroach-like robot. IEEE Control Syst Mag 19:9–14
69. Quinn RD, Nelson GM, Bachmann RJ et al (2003) Parallel complementary strategies for implementing biological principles into mobile robots. Intl J Robot Res 22(3):169–186
70. Prochazka A, Gillard D, Bennett DJ (1997) Positive force feedback control of muscles. J Neurophysiol 77:3226–3236
71. Abraham G, Nisky I, Fernandez HL et al (2012) Towards perceiving robots as humans: three handshake models face the Turing-like handshake test. IEEE Trans Haptics 5:196–207

Chapter 13
Further Discussion

13.1 Principles and Practices of Model Testing: An Overview

The use of simulation models in any field of application raises immediate concerns about the correctness of the simulation and the credibility of results obtained from it. In general terms, the processes of model testing and (in)validation are aimed at establishing whether or not a specific simulation model represents a given system with the fidelity necessary for the intended application of that model. The purpose of this section is to try to take stock of what this statement actually means, in practical terms, in the context of different types of application. Some of the findings from Chaps. 1, 2, 3, 4, 5, 6, 7, and 8 are emphasised again and information from the four case-study chapters is drawn on to provide some additional insight.

The first point to emphasise is that considerations of model quality must always reflect the purpose of the model development process, available knowledge and understanding of the real system and experimental constraints and limitations. As further knowledge is gained about the real system, whether in a scientific context or in terms of design, development and testing of a new engineering system, some aspects of the model are likely to change. Nevertheless, the fidelity should still be assessed using the same criteria throughout. Validation is not a process that is carried out only once but is an on-going procedure that is repeated many times as a model is being developed.

One relatively straightforward aspect of simulation model testing concerns internal verification and involves establishing the suitability, or otherwise, of the code and algorithms used for simulation, together with checks on the consistency of the simulation with its underlying mathematical and logical basis. The more challenging part of the testing process concerns external validation, which is concerned with critical assessment of model behaviour in relation to accepted scientific laws and principles (to establish its theoretical validity) and in terms of

D.J. Murray-Smith, *Testing and Validation of Computer Simulation Models*,
Simulation Foundations, Methods and Applications,
DOI 10.1007/978-3-319-15099-4_13

quantitative or qualitative comparisons with the behaviour of a real physical system (to establish functional, or empirical, validity). This is the aspect of model quality assessment and model testing that has been emphasised most strongly throughout this book.

The purpose of a model is always very important and comparisons of a model and the corresponding real system should always be guided by the intended application. In engineering, validation methods are likely to be mainly quantitative and can involve direct comparisons of model and system variables, as in the case study of Chap. 9 concerning the coupled tanks system. Also, in engineering applications, issues of theoretical validity are usually given priority. However, the processes of model verification and validation also apply to situations where initial knowledge about the real system is limited and where experimental investigations may be constrained by practical or ethical issues, as can arise in physiology and medicine. In the case of models with more uncertainties and imperfect experimental data, such as are found often in the biological sciences, the validation process is likely to be based on peer review and expert opinion to a greater extent, as in the sections of the case study of Chap. 12 concerning the internal organisation and structure of the neuromuscular control system. This allows functional validity to be assessed within the specific context in which the model is going to be used. Functional validity also can be linked to what has been termed "pragmatic" validity, which is concerned with assessment of the extent to which a model meets the objectives of its intended application. Such thinking can be important in areas, such as physiology, where the model may be intended simply as an aid in the testing of hypotheses about how a complex system works, as illustrated by the case study on modelling the neuromuscular system. Pragmatic validity is also important in control system design where the quality of a model of a system within a control loop has an important and direct influence on the final performance that can be achieved within the control system. Another area where pragmatic validity can be important involves fault detection systems in engineering, where early detection of faults is important but generation of false alarms is highly undesirable. Similarly, pragmatic validity is of central importance for applications where simulations are used for prediction or as an aid in decision-making processes.

Whatever the application area, the processes of validation should lead to confirmation that the model output has a level of accuracy consistent with the intended use. To carry out this process of confirmation it is essential that the accuracy requirements of the model should be established before any form of evaluation is undertaken and not as part of that evaluation process. Results of external validation are thus also best expressed in terms of the suitability of the model for a planned application, instead of as a "good" or "bad" description. Indeed, strictly speaking, one can never prove that a model is valid; a model can only be proved to be invalid, as discussed in Chaps. 2 and 7. Available evidence can be assembled to suggest that a model is suitable for an application, but more general assertions of "validity" must be avoided.

Having applied appropriate systematic procedures for model verification and validation at every stage in the development of a model, three outcomes are generally possible at the end of the modelling process. The first of these is that no model structure and associated set of parameter values can be found that is consistent with the observed behaviour of the real system and the available quantitative data. In this case the investigators must return to the model formulation stage and reconsider the problem from the beginning. The second type of outcome arises in situations where parameter values can be found that can produce model behaviour that is in agreement with the behaviour of the real system but major uncertainties exist. These uncertainties may be associated with model parameter values or with experimental measurements from the real system. In such cases the model may be judged to be of some value if it can be shown to have a sound theoretical basis and, if considered appropriate, further effort might then be devoted to gathering better experimental data through improved instrumentation and measurement techniques, or to obtaining better parameter estimates through improved experiment design and more robust methods of system identification. The concepts of identifiability and optimal design of test signals are important in this respect, as shown in the case studies concerning helicopter flight mechanics modelling and the respiratory system. The third type of outcome is the most favourable and arises when model testing leads to a situation where agreement between model predictions and experimental results is judged acceptable for the intended application, and where the model structure and parameter values are also judged to be plausible. Analysis can then establish the errors likely to arise over different parts of the operating envelope of the system. Such analysis allows limits of operation to be established for the model and, for operation within those limits, the model may then be judged to be of acceptable validity in terms of the tests carried out. The model may then be used for the planned application until new evidence is found that leads to a need for reassessment. Good documentation is vitally important at every stage so that there is a clear audit trail showing the basis of every decision leading to eventual model acceptance.

One of the complications arising in the evaluation and testing of simulation models is the fact that most models involve quantities that have to be provided by the user (e.g. as model parameter values), leading to a very large problem space. Models can also produce large numbers of output variable time histories, each with errors which may vary with time. It is important, therefore, to establish the set of output variables that are particularly important for the application. The model must always be matched to the application from the start of its development and the errors bounds must be established from the outset.

Perfect matching of all the available measured response data is never realistic and models need to match test data only to the level needed for the application. There must therefore be a trade-off between performance and robustness so that model responses match test data to an acceptable degree, while also showing satisfactory robustness to uncertainties associated with factors, such as modelling assumptions, environmental and model parameter variability or ignorance in terms of initial conditions in the real system.

At different stages of a project, a system may need to be modelled using different levels of detail and this must also be possible within the generic approach. Sub-models for component parts of the complete physical system, may therefore be needed at several levels of complexity. These may include functional forms of model for the initial stages of a project and more detailed and fully validated physically-based models as the project develops. These models, with different levels of resolution, need to be mutually calibrated in some way and the relationships between different levels of the sub-models within the generic structure must be fully understood by users. This issue of varying levels of detail within models and sub-models so that some versions of a model may be simpler in structure than others is one that has attracted attention recently and is sometimes referred to as "aggregation" [1].

13.2 Strategic Issues in Modelling and Simulation and Current Trends in Model Verification, Validation and Accreditation

The increasing computational power, falling costs and enhanced user interfaces of modern computing systems mean that significant changes are occurring in the way that people approach modelling problems, whatever the field of application. The ease with which application programs can be moved from one computing environment to another is another important factor which leads to more re-use of programs and, in a modelling context, to the development of libraries of re-usable sub-models and generic models that can be used for a range of different applications.

There are good examples, often in safety-critical application areas, such as the nuclear, aerospace, defence, marine and off-shore sectors, where rigorous model testing and formal approval schemes are routinely applied. However in other fields, model development within many engineering organisations often involves surprisingly little systematic assessment of the quality of models in terms of their useful range and limits of accuracy. Also, there may be cases where a model has a spurious justification, possibly on the grounds that it "has always been used", or is "based on well-known physical principles so must be right", or is "based on an industry standard".

In Chap. 1 mention was made of the importance of requirements specifications in the initial stages of model development and how good practices in the model development process can help to ensure project delivery within budget and on time. It is significant to note that the importance of using modelling techniques to reduce costs in large projects is also part of the philosophy of the US Defense Science Board (DSB) in the context of the DSB Model-Driven Architecture. It is stated that the proposed architecture is "…a revolutionary approach to the way modern systems are specified, designed, implemented, tested and supported. It affects all aspects of the process, and promises enormous reductions in both acquisition and lifecycle costs, as well as orders of magnitude reduction in cycle time" [2].

It can certainly be claimed, with full justification from successful projects, that within all areas of engineering design, the use of properly verified and validated physically-based models provides important evidence of performance capabilities across a wide range of operating conditions. Used appropriately, this approach can avoid the risks associated with going to the production stage with designs that turn out to be incomplete, leading to additional costs and to late delivery. The adoption of model-based methodologies for engineering systems design also provides potential benefits through the provision of improved documentation and archiving schemes that preserve important information in terms of assumptions, parameter values and test procedures in a convenient electronic format. This is extremely valuable information, much of which, in the past, has remained in the minds of the engineers involved or in their notebooks. In such situations the information was often lost forever since the relevant facts and associated judgements were usually considered too hard to locate again and re-use. Similar situations arise in other application areas, such as biology and medical applications, where modelling assumptions and limitations and other information of critical importance for correct use and interpretation of simulation results have often been inadequately recorded during model development. A systematic and well-managed approach to model specification, development, verification, validation and documentation can contribute greatly to the success of any project that involves modelling and simulation.

One illustration of the way in which some traditional barriers are being broken down can be found in the Human Physiome Project of the International Union of Physiological Sciences [3]. The term "Physiome" comes from "physio-", meaning "life" and "-ome", meaning "as a whole" and the Physiome Project is an international initiative aimed at assembling a central repository of databases and establishing links between experimentally derived information and computational models to bring all of these elements together within a single self-consistent framework. The contributors come from many laboratories in many different countries and simulation programs for the models published though the Project may be downloaded and used by others. The word "model", in the sense being used in the Physiome Project, can include everything from diagrammatic schemas that suggest relationships among elements that make up a system, to detailed quantitative descriptions. Each model accepted for the Project is an internally self-consistent summary of the available information and forms a "working hypothesis" about the system. Of special interest, in the context of this book, is the fact that predictions from the models are subjected to tests and data sets may include new results which may lead to the design of new experiments on the real system and to new models. This means that understanding of a given physiological system will, through this collaborative process of simulation, testing and comparison, be revealed in more and more detail through a step-by-step process.

Before a model is made available publicly within the Physiome Project it has to pass though what is known as a "curation pipeline". This procedure is essentially one of validation and accreditation and ensures that the simulation model is syntactically correct, semantically sound and is consistent with information provided about it in relevant publications, both in terms of the model structure and simulation results.

The recent publication of work sponsored jointly by the International Society for Pharmacoeconomics and Outcomes Research (ISPOR) and the Society for Medical Decision Making (SMDM) and carried out through the activities of their Modeling Good Research Practices Task Forces provides much information that should be of interest (see, e.g. [4–6]). The published reports from ISPOR and SMDM put emphasis on transparency as well as validation and stress the importance of people being able to see how a model is built and not just how well a model can reproduce real system behaviour. Specific recommendations are made for achieving transparency and validity through a number of "best practices" proposals. Many of these relate to documentation. Although intended primarily for those involved in healthcare system modelling, the proposals appear highly relevant for other areas of science and also for engineering research activities. Although transparency, within the context of the ISPOR-SMDM work, relates primarily to publication of scientific and medical findings and thus to the sharing of information with others, ideas relating to this apply equally to those working within a more closed community or organisation, such as a company or engineering research institute. It is believed, therefore, that important lessons could be learned from this work by people using modelling and simulation methods in many different contexts quite different from the medical and healthcare sector to which the work is directed. It has already been made clear, in earlier sections of this book, that many of the principles of model verification and validation used in modelling and simulation for defence, aerospace and other engineering applications can be applied with benefits to biological, ecological or earth sciences modelling. Similarly, recommendations from this task force in the medical field should be considered carefully by those in other areas who use simulation and modelling extensively.

Another field in which computational models and databases are becoming widely accepted is in civil engineering and construction projects. In this case it is a combination of the civil engineering and building industries and government regulations that are pushing developments forward. In the United Kingdom, adherence to Building Infrastructure Management (BIM) Level 2 requirements will be important for all construction projects funded by central government after 2016 [7]. BIM is already being used for the Crossrail project in London and will be an essential feature of the planned HS2 high-speed rail project. The link between BIM (which appears mostly concerned with visualisation, information retrieval, documentation and project cost reduction) and the dynamic models that are the subject of this book lie in the fact that BIM can provide data for use in dynamic simulations that can be of great importance in many building and construction projects. For example, BIM can automatically create input files for building energy simulation programs that are essential for energy performance analysis. Green Building XML (gbXML) is one example. This is linked to BIM efforts that are focused particularly on green building design and optimisation [8]. It provides input for several simulation modelling software systems specifically designed to be environmentally friendly.

As mentioned in Chap. 2, ideas promoted by Mr Terry Ericsen when working with the US Office of Naval Research go beyond those discussed above. He suggested that as engineering systems become more and more interconnected and more integrated in nature it will be necessary for models to take on a more fundamental role. Modelling and simulation will not be used simply as tools for analysis and, in future, the model will become the specification instead of paper documents. Ericsen envisages two types of model: requirement models and product models [9]. In his view, requirement models can be behavioural, empirical and relational but product models must be physics based and should include full details of the materials involved and the methods of manufacture of the product. Such models allow prediction of failure conditions, quantification of risk as a function of known and partly-known physics and estimation of costs as a function of materials and manufacture. This implies that if a sub-system cannot be modelled it is not understood sufficiently to allow the design to be taken forward to production. These ideas have been explored further by Famme, Gallacher et al. [10] in the context of affordability of naval shipbuilding. Their paper outlines ways in which affordability objectives can be achieved through the use of a Ship-Smart System Design (S3D) tool. They have examined likely cost savings that could be achieved in the design and construction of a destroyer-size vessel through adoption of a so-called "Performance Based Design Continuum" in which operational and human systems integration objectives are captured as dynamic models. Based on what the authors claim to be conservative estimates, a net programme cost saving could be achieved of more than 13 % which is very significant on a total cost figure of $2890M for the first ship in the class. For a total of 30 ships the life-cycle saving has been estimated at $6060M (over a period of 30 years) [10].

Savings from detailed modelling arise not only in the design phases of the project but also in commissioning and in the inevitable subsequent modernisation of the vessel in later years. Cost savings are also generated, through the availability of real-time versions of the models, in training and decision support roles on a continuing basis throughout the lifetime of the ship.

Many similarities can be seen between Ericsen's vision of the model as the specification in the context of product engineering and the concepts involved in the BIM approach for the construction industry. One can also see common ground between both of these developments that are being considered for major engineering projects and developments in biology where the Physiome Project is placing modelling methods at the centre of physiological and medical research, with models, simulations and associated databases of experimental records becoming repositories of knowledge in specific areas.

One topic that must, inevitably, be returned to in reviewing issues that are important in terms of model quality is that of fitness-for-purpose. Without a full understanding of the intended application the development of a model is, potentially, a waste of time and effort. A simple illustration of this comes from the field of automatic control engineering where models of the system to be controlled are essential for most methods of control system design. Models are also very important in terms of their use within failure-detection and isolation systems but the requirements in terms of the models for these two types of application are very different [11].

For control system design a simplified model of relatively low order is often best since it leads to controller designs which also are relatively simple and also of low order. The fact that the presence of feedback can reduce the effect of nonlinearities and parameter variations on the overall closed-loop system design is also helpful. Good closed-loop system designs usually result, provided appropriate design techniques are used and the models are sufficiently accurate in key parts of the frequency range being considered. In the context of a single-input single-output system this is especially important in the part of the frequency range where the magnitude ratio is tending to unity and the phase shift is tending towards −180°. Provided we have an accurate model within that part of the frequency range, it is possible for the closed-loop system to meet given stability requirements and also satisfy the specification in terms of steady state errors and robustness to external disturbances. The detailed performance of the controller can then be checked further by doing additional tests using a more complex version of the model which may be of higher order than the model used for design and may include nonlinearities.

The situation is different when models are being developed for use within failure detection and isolation systems. Model-based failure detection methods involve comparisons between measured responses and responses for the same operating conditions and input. Problems arise both due to model inaccuracy and to noise in measured response data. Although methods are available which can deal with sensor noise using averaging techniques which reduce the effects of noise without interfering with the failure signature, model errors remain a difficulty and models which include as many features of the real system as possible have potential advantages. Successful fault detection depends on the use of appropriate "decision thresholds" in order to maximise detection of faults and at the same time minimise false alarms. Modelling errors are clearly important, but so is the need for real-time operation of the simulation model being used within the fault detection system. A compromise may therefore have to be reached in some applications between the model accuracy and speed.

13.3 Research and Development Opportunities in the Field of Model Quality Assessment

In the autumn of 2002 a group of almost 200 people from Belgium, Canada, France, Germany, the United Kingdom and the United States met in the USA for a "Workshop on Foundations for Modeling and Simulation (M&S) Verification and Validation (V&V) in the 21st Century" [12]. This is now perhaps better known as "Foundations '02" and the proceedings papers for that event describe very fully the verification and validation practices of a range of organisations (mostly large) in all of the countries involved. The conclusions of that workshop make interesting reading, as do the recommendations in terms of areas where understanding, at that time, was not as complete as was thought desirable. A paper which provides a useful summary of the Foundations '02 Workshop has been published by Dale Pace and in this he highlights a number of management and research challenges that were identified during the Workshop [13]. Although the Workshop took place over a decade ago, most of the challenges identified in 2002 are still very relevant. For example, there are clearly many management challenges. One of these is concerned with the difficulties of qualitative assessment involving face validation methods and the need for objectivity and consistency. A second challenge is the need to tap into available expertise in terms of formal methods so that appropriate "lightweight" versions of proven methods from other fields can be assessed in the context of simulation modelling. However, one of the biggest challenges concerns the issue of costs and resources and little information on this is currently available. A number of other challenges relating to research were also identified at the Foundations '02 Workshop, including the need for new and more rigorous methods for making inferences about simulation results, advances to deal with the problems of adaptive systems, new theory to allow issues of aggregation to be handled more effectively and new methods for the representation of human behaviour within simulation models.

13.4 Educational Issues

University students in the physical sciences and engineering are usually introduced to mathematical modelling and encounter computer-based modelling and simulation methods early in their university education. However, topics relating to model management are completely neglected in many courses and most students seldom have to give serious thought to what constitutes a simulation model that is fit for purpose. Indeed, all issues of model quality are often glossed over in a superficial fashion and the teaching often stops with the formulation of equations from physical laws and principles, or with linear models obtained experimentally by system identification and parameter estimation methods. Also, students often do not make the vitally important link between design success and model quality and fail to appreciate that correction at a late stage in a project for model inadequacies introduced much earlier can lead to major additional costs and delays in completion. In the biological sciences relatively few courses appear to put any emphasis on system modelling and, although exposed to statistical concepts from an early stage, few students are likely to have any understanding of the potential importance of computer simulation methods in their field.

Degree courses in all areas of science and engineering must cross traditional boundaries to a greater extent and include realistic practical exercises involving modelling and simulation methods. Ideally, these exercises should push students beyond their normal comfort zones and involve them in investigations that are more open-ended, especially for courses at first-degree level. Student exposure to modelling should range from initial scoping of problems, involving back-of-the-envelope style exercises, through combined experimental and simulation studies of laboratory-scale hardware to group project activities. In engineering, for example, this might involve, in the later years, design projects that could involve techniques such as virtual prototyping, hardware-in-the-loop simulation and embedded systems. The modelling and simulation aspects of such project work should, where appropriate, provide an introduction to the use of model libraries and generic models. Good management and effective use of modelling and simulation tools should be emphasised in courses on modelling and simulation and should form an essential element of project work. Students should be aware of model quality issues from an early stage in their training and this means that they must be exposed to modelling and simulation ideas repeatedly and creatively from the earliest stages of their degree programmes.

13.5 Final Remarks

One of the essential messages from the case studies chapters is that, in assessing the credibility of results from a physically-based simulation model, it is not sufficient to apply only one or two tests or even one or two validation methods. A broad view must be taken of model testing and validation procedures and every available piece of evidence must be fully taken into account in assessing the fidelity of a model. It is never appropriate to neglect some inconvenient fact that does not fit with the other information gathered about the system and the model. Every item of available information must be used. Full account must also be taken of expected errors and uncertainties, both within the model and within the data and measurements from the real system.

Model validation is sometimes compared with the legal processes of establishing guilt or innocence. In most legal systems the accused is either declared innocent or guilty, beyond any reasonable doubt. Similarly, whether the validation process is based on quantitative measures or face validation, the required outcome must involve accepting a model as suitable for the intended application, beyond reasonable doubt, or rejecting it as being unsuitable. As with all analogies, there are differences between these situations and the validation process also provides information about the range of conditions over which a given model is a useful representation. It must also be remembered that a model can (and should) be re-tested when new evidence comes from the corresponding real system that was not available when the model was originally accepted. This is equivalent to a re-trial when new, legally-admissible, evidence becomes available.

Model validation should always be viewed as a process rather than an end result. That process should, above all, ensure that only models that can be shown to be appropriate, with full supporting evidence, are brought into use for their specific intended application. Building confidence in a model, whether it is intended for use in design, in decision making, or as a tool in scientific research, is always an iterative procedure that requires a careful, well-documented and rigorous approach in which quality is the long-term goal. Although many techniques are available that can be usefully applied, current practice in model testing and evaluation falls well short of what could be achieved in many application areas and in many organisations. It is hoped that this book may help, to some extent, to make those who develop simulation models and also those who use them more aware of the need for the rigorous application of currently-available good-practice procedures. It is hoped that it may also stimulate further work on improved methods, procedures and recognised standards in terms of model testing and evaluation. There is much scope for improvement and the role of those involved in educating the next generation of engineers, scientists and applied mathematicians is particularly important.

References

1. Bigelow JH, Davis PK (2003) Implications for model validation of multiresolution, multiperspective modelling (MRMPM) and exploratory analysis, Report MR-1750-AF, RAND Corp, Santa Monica, CA
2. Anonymous (2006) Defense Science Board, Model Driven Architecture (MDA), DSB Report, Appendix 1, p 85, Defense Science Board, Washington DC
3. Anonymous, About the IUPS Physiome Project. www.physiomeproject.org/about/the-virtual-physiological-human. Accessed 30 May 2015
4. Caro JJ, Briggs AH, Siebert U et al (2012) Modeling good research practices – overview: a report of the ISPOR-SMDM Modelling Good Research Practices Task Force -1. Med Dec Making 32(5):667–677
5. Briggs AH, Weinstein MC, Fenwick EAL et al (2012) Model parameter estimation and uncertainty: a report of the ISPOR-SMDM Modelling Good Research Practices Task Force – 6. Med Dec Making 32(5):722–735
6. Eddy DM, Hollingsworth W, Caro JJ et al (2012) Model transparency and validation: a report of the ISPOR-SMDM Modeling Good Research Practices Task Force–7. Med Dec Making 32 (5):733–743
7. Building Information Modelling (BIM) Task Group website. www.bimtaskgroup.org. Accessed 30 May 2015
8. Green Building XML schema website www.gbxml.org. Accessed 30 May 2015
9. Ericsen T (2005) Physics based design, the future of modeling and simulation. Acta Polytech 45(4):59–64
10. Famme JB, Gallagher C, Raitch T (2009) Performance based design for fleet affordability. Nav Eng J 12(4):117–132. doi:10.1111/j.1559-3581.2009.00233.x
11. Horak DT (1989) Experimental estimation of modelling errors in dynamic systems. J Guid Control Dyn 12(5):653–658
12. Foundations for V&V in the 21st Century Workshop (2002), (Foundations '02), John Hopkins University, Applied Physics Laboratory, Laurel, Maryland, 22–23 October 2002. http://www.docstoc.com/docs/92047381/A-model-and-simulation-verification-and-validation-_V_V_-workshop. Accessed 12 June 2015
13. Pace DK (2004) Modeling and simulation verification and validation challenges. Johns Hopkins APL Tech Dig 25(2):163–172

Index

D.J. Murray-Smith, *Testing and Validation of Computer Simulation Models*,
Simulation Foundations, Methods and Applications,
DOI 10.1007/978-3-319-15099-4

Printed in the United States
By Bookmasters